RISKY WORK ENVIRONMENTS

Risky Work Environments
Reappraising Human Work Within Fallible Systems

Edited by

CHRISTINE OWEN
University of Tasmania, Australia

PASCAL BÉGUIN
Director of Research at the French National Institute for Agricultural Research (INRA), France

&

GER WACKERS
University of Maastricht, Netherlands

ASHGATE

© Christine Owen, Pascal Béguin and Ger Wackers 2009

All rights reserved. No part of this publication may be reproduced, stored in a retrieval system or transmitted in any form or by any means, electronic, mechanical, photocopying, recording or otherwise without the prior permission of the publisher.

Christine Owen, Pascal Béguin and Ger Wackers has asserted their moral right under the Copyright, Designs and Patents Act, 1988, to be identified as the editors of this work.

Published by
Ashgate Publishing Limited
Wey Court East
Union Road
Farnham
Surrey, GU9 7PT
England

Ashgate Publishing Company
Suite 420
101 Cherry Street
Burlington
VT 05401-4405
USA

www.ashgate.com

British Library Cataloguing in Publication Data
Risky work environments : reappraising human work within
   fallible systems
   1. Industrial safety – Psychological aspects 2. Human
   engineering 3. Hazardous occupations
   I. Owen, Christine II. Beguin, Pascal III. Wackers, G. L.
   (Ger L.)
   363.1'1

Library of Congress Cataloging-in-Publication Data
Owen, Christine.
  Risky work environments : reappraising human work within fallible systems / by Christine Owen, Pascal Beguin and Ger Wackers.
     p. cm.
   ISBN 978-0-7546-7609-6
   1. Work environment. 2. Risk assessment. I. Beguin, Pascal. II. Wackers, G. L. (Ger L.) III. Title.

HD7261.O946 2009
363.11--dc22

2008045375

Printed and bound in Great Britain by
TJ International Ltd, Padstow, Cornwall

Contents

List of Figures   *vii*
List of Tables   *iv*
About the Contributors   *xi*

1   Introduction: Shifting the Focus to Human Work within Complex Socio-technical Systems   1
Pascal Béguin, Christine Owen and Ger Wackers

PART I   **IDENTIFYING SYSTEM VULNERABILITIES WITHIN INCIDENT AND ACCIDENT ANALYSIS**   **11**
Ger Wackers

2   Learning from Accidents: Analysis of Normal Practices   17
Lenna Norros and Maaria Nuutinen

3   Derailed Decisions: The Evolution of Vulnerability on a Norwegian Railway Line   53
Ragnar Rosness

4   Offshore Vulnerability: The Limits of Design and the Ubiquity of the Recursive Process   81
Ger Wackers

PART II   **ACCOMPLISHING RELIABILITY WITHIN FALLIBLE SYSTEMS**   **99**
Christine Owen

5   Channelling Erratic Flows of Action: Life in the Neonatal Intensive Care Unit   105
Jessica Mesman

6   How do Individual Operators Contribute to the Reliability of Collective Activity? A French Medical Emergency Centre   129
Jacques Marc and Janine Rogalski

PART III	ENHANCING WORK PRACTICES WITHIN RISKY ENVIRONMENTS *Pascal Béguin*	**149**
7	When Users and Designers Meet Each Other in the Design Process *Pascal Béguin*	153
8	Near-misses and Mistakes in Risky Work: An Exploration of Work Practices in High-3 Environments *Christine Owen*	173
9	Conclusion: Towards Developmental Work within Complex and Fallible Systems *Christine Owen*	197
Index		*207*

List of Figures

Figure 2.1	The path of the accident investigation and results of the phases	28
Figure 2.2	The pilots' performance three groups (A–C)	37
Figure 2.3	Different views of navigation methods	39
Figure 2.4	Interpretation of the differences in the rated piloting practices (performance) with the aid of historical types of navigation	40
Figure 2.5	Piloting as an activity system	41
Figure 3.1	The situation immediately after the northbound train had left from track 1 at Rudstad, forcing open the switch at the northern exit of the station	57
Figure 3.2	Causal and contributory factors in the Åsta accident	67
Figure 3.3	A second representation of the Åsta accident	75
Figure 8.1	Intersection of temporal, complex and interdependent nature of High-3 work practices	178

List of Tables

Table 2.1	The core task model of sea piloting	30
Table 2.2	Issues that complicated the conditions for safe piloting in the different cases	35
Table 3.1	Traffic control on the Røros line at the time of the Åsta accident compared to manual traffic control and traffic control on main lines	60
Table 7.1	Comparative evolution of operators' gazes (in percentage of total session observation time) at the prototype and at previously available thermometers during the first, third, and sixth working hypothesis sessions	164
Table 8.1	Dimensions of High-3 work, their elements and the resources employed to mediate risky work practice in air traffic control	179
Table 9.1	Theoretical resources for developmental work activity	199

About the Contributors

Christine Owen is a senior researcher investigating communication, coordination and collaborative practices in high-technology, high-intensity, high-reliability work organisations. She has conducted research in aviation, and emergency medicine and emergency management environments. She has a particular interest in theories of activity and how developmental work practices may be enabled. She is currently Assistant Dean of the Faculty of Education at the University of Tasmania, Australia and National Research Program Leader for Education, Training and Communication in the Australasian Bushfire Cooperative Research Centre.

Pascal Béguin is director of research in ergonomics at the French National Institute of Agricultural Research (INRA). He has conducted researches in engineering companies, aviation, chemical industry, medicine (oncology) and agriculture (sustained development). Pascal Béguin represents the French tradition of interventionist studies of design in work settings. He has a longstanding interest and experience in transformative approaches to participatory approaches in design. He has published extensively on the notion of instrumental genesis and its implications for the organization of design processes. Pascal Béguin is a member of the French Institute for Research on Innovation and Sciences (IFRIS, University of Paris-east) and executive editor of the open access journal @ctivités.

Ger Wackers is associate professor at the Department of Applied Sciences, University College Narvik, Norway. He was originally trained, in the Netherlands, as a medical laboratory technician and as a medical doctor. Subsequently he moved into science and technology studies (STS). He holds a PhD in STS from Maastricht University. Until recently he was associated with the Department of Technology and Society Studies of the Faculty of Arts and Culture of Maastricht University, where he worked on a research project on the vulnerability and reliability of complex technological systems, particularly in the offshore industry in the North Sea. He continues his offshore industry and vulnerability related research at the University College in Narvik which is focused on issues of cold climate technology and work in the High North. He has an interest in transdisciplinary approaches in STS research and has been writing with engineers on issues of reliability of complex technological systems.

Ragnar Rosness is senior scientist at the Department of Safety and Reliability at SINTEF in Trondheim, Norway. He holds an M.Sc. in psychology and a PhD in industrial engineering. He conducts research mainly in the petroleum industry and the transportation sector. His research interests focus on the links between organizing, decision-making and vulnerability.

Leena Norros is a research professor in human factors at the Technical Research Centre of Finland where she heads a research group in Systems Usability. Her area of expertise is work psychology and cognitive ergonomics of complex industrial systems. She studied psychology at Helsinki University and took a PhD in work psychology at the Technical University of Dresden and at Helsinki University. Currently she acts as docent in work psychology at Helsinki University. She has developed an original ecological research approach to human activity, in which she makes use of the cultural historical theory of activity and the pragmatist conception of habit. In her practice-related research work she combines the Nordic developmental work research with the European human factors tradition.

Maaria Nuutinen is a work psychologist studying problems of cognitive ergonomics in complex industrial systems at the Technical Research Center of Finland. She completed her master thesis on driver behaviour and graduated from Helsinki University in 1997. She is currently researching professional competence and identity themes.

Jessica Mesman is assistant professor at the Department of Technology and Society Studies, Maastricht University. She was trained as a professional nurse and obtained a degree in the Philosophy of Health Sciences. She holds a PhD in Science and Technology Studies based on an international ethnographic study on medical-ethical uncertainty on an intensive care unit for newborns (NICU) in the Netherlands and the United States. She is currently doing an ethnography of knowledge practices to explicate the hidden competencies and informal structures that are part of systems of safety in critical care medicine.

Jacques Marc is researcher at the French National Institut for Research on Safety (INRS) in the Département Homme au travail Laboratoire. He obtained a PhD in psychology of work at University Paris 8. His research interests focus on human individual and collective activity in open dynamic environments.

Janine Rogalski is involved in research at the University of Paris 8, where she is a senior researcher. She holds degrees in the didactics of mathematics and in psychology. Her research interests include human individual and collective activity and competence development in open dynamic environments. She has conducted research in the empirical domains of aviation/air traffic control and of fire fighting and disaster management.

Chapter 1

Introduction: Shifting the Focus to Human Work within Complex Socio-technical Systems

Pascal Béguin, Christine Owen and Ger Wackers

It is likely that most researchers and practitioners operating within (that is, 'with' and 'in') complex technological systems would agree with the observation that humans contribute positively to mitigating risk. Indeed, there is a growing consensus that researching error alone cannot help us to fully understand how reliability in systems is accomplished. The contributors in this book show how such reliability is created in the ways in which humans work within their systems, positively taking into account the fallibility of those systems. This book is, therefore, about illustrating the ways in which reliability is accomplished most of the time, despite imperfect systems, technologies and organizational procedures.

> The core question of this book is:
>
> *How do workers operating within risky environments and in imperfect systems positively work to mitigate those risks and what problems do they encounter in so doing?*

The authors of this book contribute to how we can better understand safety in systems. They contribute to this theoretical development through their focus on the role of human work in the design of complex technological systems. They do this first, by analysing the positive role of human actions *with* those systems as well as *in* those systems. Second, they do so through developing better understandings of the ways in which humans accomplish work in those environments, despite their own or systemic limitations, even in the case of failure. In this way the contributors move beyond typical error analysis toward analyses of how human operators positively act to mitigate risk in everyday operations.

In this book, readers will find a rich body of data from empirical investigations in a range of industrial sectors, and will meet:

- ship pilots navigating a diversity of seagoing ships through the channels and between the islands in a coastal archipelago;
- rail traffic controllers trying but failing to avoid a frontal collision between two trains;

- engineers working within the multi-billion dollar offshore mining industry and whose work we can better understand within the context of broader socio-economic and political influences;
- medical doctors and nurses in neonatal intensive care units who must find treatment trajectories in the face of uncertainty about what is wrong with the seriously ill and premature newborns in their care, and about what the long-term results may be;
- staff who work in an emergency medical communications centre and who constantly strive to integrate potentially conflicting internal and external goals and associated risks to ensure 'acceptable' operational performance;
- designers involved in developing artefacts to avoid runaway chemical spills, who face the challenge of better understanding the ways in which the tools they develop are redesigned in practice by users;
- air traffic controllers who juggle the temporal demands of complex work within the constraints and opportunities of technologically mediated interdependent work.

Two arguments are central to developing a more holistic understanding of human work in complex systems. The first is that systems are fallible and that we need to move the focus away from human error to the dynamics of systems and vulnerabilities. The second is placing human work practices at the centre of theory building as a core to achieving this understanding. These two arguments are at the heart of the discussions between the authors of this book.

From Error ... to Working within Fallible Systems

The history of human factors research and its account of error and accidents has its own developmental trajectory. In the early days of civil aviation, for example, the main focus on post-accident accounts of error was individual human blame. Realizing this accomplished little, the focus shifted from one of individual blame to post-hoc analyses of accidents that attempted to be more systemic (Maurino 1995).

However, these accounts also came under criticism for their sometimes overly simplistic accounts of a linear cause-effect approach. And once more, while they offered a more satisfying account of what accounted for the error, they remained, for the most part, overly focused on what went wrong, typically in hindsight (Woods and Cook 1999).

One of the key contributions of this phase of attempting to understand the causes of accidents more holistically was the attention given to what James Reason (1990) called 'latent factors' (for example, training, organizational policies, organizational culture). These insights led to the rise of programs associated with crew resource management, where the focus was on strategies to help crew members draw on the communicative resources available within the team and not just technical abilities, to delegate tasks and assign responsibilities, facilitate communication and cross-checking (Helmreich and Foushee 1993).

Subsequent iniatives to improve the safety of high reliability organizations led to the development of various iterations of crew resource management training, specifically designed to decrease the probability of errors occurring before they had a chance to impact (Helmreich and Weiner 1993). Such studies shed insight into the different types of errors made in practice, why they were made and the conditions under which types of error are made (for example, Weigmann and Shappell 1997). Yet still the focus was on error, which was fortunately becoming even more rare in high reliability environments (see Amalberti 2001; Reason 1997, 2004).

These studies migrated into accounts of safety through investigating threats and risks in normal everyday operations. While the focus of these studies (see, for example, Hawkins 1993; Johnston, MacDonald and Fuller 1997) have moved from investigating accidents to addressing how operators mitigate threats, and associated risks attention still remains on mistakes and abnormality.

The pervasiveness of error analyses

Error persists as a focus for at least three reasons. One possible explanation is that there are multiple purposes that the error analysis and conclusion will serve. There is, for example, a need to identify and attribute responsibilities for the incident or the accident after the fact. This has to do with accountability. A human error-cause is linked to the idea that only humans can be held accountable. A second reason has to do with learning. Given the enormous consequences of failure in complex systems, what can be learned so that similar mistakes might be avoided in the future? The third reason is also deeply anchored in our culture. We need to have faith in the systems we create and the technologies we use. The notion of an 'error' allows us to localize and isolate a specific deviation from 'designed' performance criteria, a deviation that can be repaired. Recognizing (in the sense of admitting) and repairing the 'error' is important in re-establishing trust in particular technologies that have failed but also, in a more general sense, to maintain or re-establish trust in the complex technological system on which our technological culture has become so dependent. Many different types of analysis in this book will focus attention on what is problematic in each of these views.

Toward Human-centred Work Practices

For some time there has been a growing consensus that error-focused analyses of human work in complex systems is inadequate and that new research methodologies and approaches are needed (Dekker 2006; Hollnagel 2005; Hollnagel and Woods 1999; Rasmussen 1990, 2000, 2003; Reason 1990; Vicente 1999; Weick 1999).

This has occurred for two reasons. As indicated, given that catastrophic accidents were rare, it was questionable what contribution they could offer to understanding what happens in workplaces most of the time. Second, there was an increasing dissatisfaction with such approaches that were based on understanding what happens in complex systems only through analysis of what happens when things go wrong and an increasing desire to move beyond 'deficit' models (e.g., Flach 1999; Hollnagel 2005). So, within the human factors and safety science literature is increasing dissatisfaction with using the study of accidents and error as the only means by which we can understand resilience in complex systems. However, although there are many exhortations in the literature to 'shift the focus beyond error' (Hollnagel 1993, 2005; Hollnagel and Woods 1999), such analyses have often still remained centred on the loss of comprehension or the failure. This book takes for granted the importance of developing holistic ways of understanding everyday human work in complex environments (Rochlin 1999; Rasmussen 1990, 2000; Reason 1990).

Redefining the problem of working within risky environments

Our claim here is that the everyday work environment is inherently imperfect whether or not the system is in a degraded mode state. The contributions in this book begin to develop this shift toward a more human-centred focus on work within fallible systems by illustrating the ways in which humans contribute positively to mitigating risk, despite fallible technologies, unrealistic rules and procedures in organizations and systems that are inherently imperfect, by their very nature.

However, it is also acknowledged that a positive approach to examining work within risky environments is not completely new. In 1967, Perrow proposed a classification of tasks based on their variability, and suggested that a high variability with a low predictability required a flexible and decentralized work organization in which the worker must build understanding and solve problems as and when they appeared. Rasmussen (2000) noted that, as routine tasks are progressively automated, the importance given to activity in situations and to user intelligence has constantly grown. Through their activities within risky environments and within imperfect systems, humans mitigate risk. However, a key challenge yet to be fully addressed in the literature is: *How are we to better understand and support professional practices within fallible systems?*

The book urges readers to recognize that often technologies are fallible. Indeed, the chapters in this book provide strong empirical evidence for the argument that reliable performance occurs in complex environments *despite* often conflicting constraints imposed by divergent organizational goals with respect to safety, availability, profitability and regularity. Devices or rules and procedures are not sufficient for achieving successful, safe and reliable work. Moreover, safety cannot be achieved by strict proceduralization or any other form of standardization in complex systems.

The contributors use a diverse range of theoretical perspectives to examine human work within complex technological systems by bringing together empirical research. They draw on foundational studies in human factors analysis such as cognitive ergonomics and complexity theory as well as others, such as Cultural Historical Activity Theory (CHAT) and approaches from the field of science and technology studies (STS), specifically, actor-network theory. There are a number of French-speaking authors in the volume. The French-speaking ergonomic community has a rich history and has elaborated on French-language concepts not well known to English-speaking audiences.

Writing from different disciplines and theoretical perspectives, the contributors in the book highlight differing questions, levels of analysis, problems and solutions. But in spite of their diversity, they share the same aim: drawing attention to individual, collective, systemic and technological resources that can be used to accomplish successful work in risky environments. Thus the diversity of approaches taken in the book provides the reader with a set of theoretical resources with which to analyze human work in complex environments.

For example, Marc and Rogalski (Chapter 6) analyze why, despite a high level of errors made by individuals in an emergency communications centre, at the level of the collective, safety remains high. If theoretical development relies only on developing an understanding of the health of systems through the lens of error analysis, then these kinds of accomplishments will remain unidentified.

Several analyses of serious accidents are also outlined in this book. However, rather than identifying a culprit, contributors analyze the socially constructed and contextual nature of error and accident inquiries. The importance of these studies is that they draw into their analysis the paradigms, assumptions and political pressures that are inherent in the work of accident investigation, something not often acknowledged in traditional forms of error analysis.

Despite their differing theoretical orientations, the focus for the contributors here is to define the problems to be solved starting from the human work that is undertaken. In analysing how humans (in and through the systems they create) mitigate risks, this focus begins to articulate strategies for developing, designing and expanding individual, collective, organizational and technological resources for work in risky environments.

There are paradoxes faced by workers operating within fallible systems. Double-binds are present in much work practice, for example, between the need for safety on the one hand and the need for efficiency on the other; or between working within the bounds of procedures, and needing to improvise within action. In part, such paradoxes stem from the managerial assumptions about 'bounded rationality' (Shotter 1983), that is, the problem of uncertainty in work can be reduced by encoding information into organizational structure. The more completely the structure of an organization can provide needed information, the less uncertainty there will be: the human operator just needs to follow the course of known rational action.

As already indicated, a key theme in this book is that rules and procedures are not sufficient for the successful achievement of safe and reliable work. Here, we analyze the ways in which people make decisions in the interface between what is procedure and what is sometimes required as improvised action. A key question then for the contributions is: *what light can they shed on the balance between needed prescription and proceduralization, and initiative that can be encouraged and enabled? What needs to be decided and what can be left to the decision-maker to be created in action?*

One solution to such paradoxes can be found in the process of problem 'visibilization' in organization (Engeström 1999; Rasmussen 2000). The process of making the problem visible is important during analysis and needs to be used in this context (Leplat 2005; Hollnagel 2005). However, organizations are paradoxical places. Making problems visible inside an organization needs to be undertaken carefully.

Under some conditions there are systemic contradictions in organizations where it is simply not acceptable to make visible the violation of a procedure. This is because of the close relationship between knowledge and power. Problem visibilization is knowledge, intertwined within power relations. If people feel that they will be exposed and vulnerable (because of organizational or public sanctions) for visibilizing violations in work practice, then doing so will be resisted. Such power relations are also socio-culturally constructed. If a problem made visible is at odds with the organizational culture of a particular group, then this too will be resisted by the workers themselves.

Therefore, although making problems visible is a resource, it too can be problematic. Intelligence for system development can be gained from doing so. However, if conditions do not allow for making problems visible then system development will not occur. Interventionists operating in complex environments need to take special care to evaluate the possible tensions and systemic or structural contradictions that are present to enable violations to be named. Contradictions (that is, tensions in systems that are often contradictory) can be positive or negative in terms of opportunities for system development. A key question for researchers of human work within fallible systems is *what are the conditions that lead to enabling visibilization of contradictions to provide opportunities for positive development of the system?*

Enhancing Development of the Positive Role Humans Play in Fallible Systems

A key insight from a number of this book's contributors is how people actively repair and make good imperfect systems. Indeed, that systems are imperfect and fallible is easily demonstrated. Often work is accomplished efficiently because a judgment is made that a procedure can either be modified or ignored. Gaining efficiencies is often dependent on this paradoxical practice: one only needs to point to how systems grind to a halt or break down when workers collectively 'work to rule'.

In this volume there are many examples of how people negotiate within such fallible systems to create resilience and reliability. Marc and Rogalski (Chapter 6) draw attention to how emergency communication dispatch operators work collectively to cross-check and where necessary modify actions of colleagues to ensure mistakes are avoided. Other contributors have analyzed how humans work within poorly designed and flawed systems, and where, for the most part, successful work is accomplished (see, for example, Chapters 5 and 7). In contrast, other chapters provide examples of what happens when system vulnerability drifts too far and when operators, despite their best efforts, fail to realize successful work and dangerous failure results instead (see, for example, Chapters 2, 3 and 4).

If we accept that people operate positively and meaningfully in relation to the context at hand, then it is important to expand the repertoire of resources they will have available to them in their work environments. But will doing so be enough? Or given that this is the main intention of much design work, for example, why is this not enough? These questions can be usefully related to the proposals of Vicente (1999). He argues that it is not enough to model observed behaviour, but rather what is needed is a new generation of 'formative' models. The key concept underpinning these models is the observation that it is impossible to anticipate and to solve every event that can arise in risky environments through procedure, artefact or organization. Rather, anticipation can be defined as a resource rather than a dictate, which helps workers to create and develop their own best practices. So anticipation defines boundaries on action, within which workers develop their own actions.

Not all boundaries are defined in rules or procedures, nor are they about violation of work practice. A number of authors (for example, Rasmussen 2000, 2003; Rasmussen and Sveding 2000; Reason 1990) have drawn attention to the conflicting and sometimes inconsistent boundaries of profitability on the one hand and safety on the other. We agree with Rasmussen (2000) that boundaries must be made visible so that understanding their conflicting and inconsistent elements can also be used as a resource to develop mutually agreed and managed practices. This visibilizing of boundaries helps workers to stay within acceptable limits of safety. What appears in many of this book's chapters is the nature of organizational and artefactual dynamics or 'boundaries' that push workers and systems beyond their limits.

A formative approach, however, must not be reduced to defining action on the basis of what is previously known or decided (by designer, management or trainers). Knowledge and practices have to be developed and expanded by the workers themselves, through improvisation and experience. The importance of this last argument is probably the most central to the messages contained in this book.

This book is organized into three parts:

1. Part I: Identifying system vulnerabilities within incident and accident analysis.
2. Part II: Accomplishing reliability within fallible systems.
3. Part III: Enhancing work practices within risky environments.

Part I utilizes three contributions that take us directly into the world of catastrophic accidents and the work practices of those responsible for their investigation. By re-investigating critical incidents and accidents with severe consequences, it is not our intention to second-guess the work of safety inspectors and others responsible for risk assessment and accident analysis, but rather to contribute alternative ways of thinking about accidents. The authors carry out fine-grained (that is, moment-by-moment) analysis, working from perspectives and with conceptual tool-kits that are different from the formal methods of investigation and conventional modes of analysis typically employed in investigations and public inquiries. Although the analyses in this book imply criticism of traditional processes used in accident investigation, they do not rest in criticism but aim to generate resources for practitioners in safety inspections, risk assessment and risk management. Thus, one question addressed in this section is *what can we learn from these investigations for the enhancement of the whole system?*

Part II comprises two contributions that analyze how humans accomplish reliable performance most of the time despite system vulnerabilities: disturbances, high degrees of uncertainty and unpredictable events. The authors analyze the kinds of adaptive strategies, improvisations, negotiations, conflict between logics, and compromises enacted in order to repair disturbances and achieve performative closure. The authors address questions relating to how workers accomplish reliability by employing the plasticity available within their systems, how they adapt or negotiate their own individual and collective course of actions, or how they make use of artefacts and the work organization. Thus, one question addressed in this Part is *what can we learn from these adaptive processes situated at the level of human action within complex technological systems?*

In Part III the question is not about adaptive processes as such, but the development and enhancement of these adaptive processes over longer periods of time than that contained within the analysis of action of a case. Three contributions are presented, relating to the development of professional practices. Together, the three chapters discuss different ways of knowing. Different ways of knowing are organized contextually (for example, during the design process or daily work activities). Different ways of knowing are also represented within formal and tacit understandings, between, for example, the shared professional worlds of work and designers; or between the individual and the collective and distributed over artefacts. In this Part we ask *how can these different ways of knowing and learning be used as a resource to enhance professional practices within risky environments?*

One of the main arguments of this book is that it is important to articulate the individual, collective and systemic resources utilized by the workforce (and available within the systems they create) who are faced with risks. The capability of a organization's employees and that which is embedded within its systems are among an organization's more important assets.

The contributions in this book present its readers (who we anticipate are researchers, practitioners, designers, managers) with some difficult questions that they themselves will need to face while working to achieve outcomes within their own systems. In developing interventions within their practice, they will be confronted with many of the challenges raised in this book. These challenges will not be easily resolved. Working within fallible systems requires concrete problem-solving and working with dilemmas in practice that are complex, paradoxical and constantly changing. The contributions herein signal an important shift and articulation of future agendas for research and practitioner development.

References

Amalberti, R. (2001), 'The Paradoxes of Almost Totally Safe Transportation Systems' *Safety Science* 37: 109–26.

Dekker, S. (2006), *The Field Guide to Understanding Human Error* (Aldershot: Ashgate).

Engeström, Y. (1999), 'Expansive Visibilization of Work: An Activity Theoretical Perspective', *Computer Supported Cooperative Work* 8(3): 63–93.

Flach, J. (1999), 'Beyond Error: The Language of Coordination and Stability', in Carterette, E. and Friedman, M. (eds.), *Handbook of Perception and Cognition: Human Performance and Ergonomics*, 2nd edition (San Diego: Academic Press).

Hawkins, F.R. (1993), *Human Factors in Flight* (Aldershot: Ashgate).

Helmreich, R.L. and Foushee, H.C. (1993), 'Why Crew Resource Management? Empirical and Theoretical Bases of Human Factors Training in Aviation', in Weiner, E.L. and Kanki, B. (eds.) (1993), *Cockpit Resource Management* (San Diego: Academic Press).

Helmreich, R.L. and Weiner, E.L. et al. (1993), 'The Future of Crew Resource Management in the Cockpit and Elsewhere', in Weiner, E.L and Kanki, B. (eds.) (1993), *Cockpit Resource Management* (San Diego: Academic Press).

Hollnagel, E. (1993), *Human Reliability Analysis: Context and Control* (London: Academic Press).

Hollnagel, E. (2000), 'Analysis and Prediction of Failures in Complex Systems: Models and Methods', in Elzer, P.F., et al. (eds.) (2000), *Human Error and System Design and Management* (Verlag: Springer).

Hollnagel, E. (2005), 'Learning from Failure: A Joint Cognitive Systems Perspective', in Wilson, J.R. and Corlett, N. (eds.) (2005), *Evaluation of Human Work*, 3rd edition (Boca Raton: Taylor and Francis).

Hollnagel E. and Woods, D.D. (1999), 'Cognitive Systems Engineering: New Wine in New Bottles', *International Journal of Human-Computer Studies* 51: 339–56.

Johnston, N., MacDonald, N. and Fuller, R. (1997), *Aviation Psychology in Practice* (Aldershot: Ashgate).

Leplat, J. (2005), 'De l'activité contrôlée à l'activité erronée', Personal communication.

Maurino, D.E. et al. (1995), *Beyond Aviation Human Factors* (Hants: Averbury Technical).

Perrow, C. (1967), *Normal Accidents: Living with High Risk Technologies* (New York: Basic Books).

Rasmussen, J. (1990), 'Learning from Experience? How? Some Research Issues in Industrial Risk Management', in Leplat, J., and Terssac, de G. (eds.) (1990), *Les facteurs humains de la fiabilite dans les systemes complexes* (Toulouse: Octares Entreprise).

Rasmussen, J. (2000), 'Human Factors in a Dynamic Information Society: Where are we Heading?', *Ergonomics* 43(7): 869–79.

Rasmussen, J. (2003), 'The Role of Error in Organizing Behaviour', *Quality and Safety in Health Care* 12: 337–83.

Rasmussen, J. and Svedung, I. (2000), *Proactive Risk Management in a Dynamic Society* (Swedish Rescue Services).

Reason, J. (1990), *Human Error* (Cambridge: Cambridge University Press).

Rochlin, G.I. (1999), 'Safe Operation as a Social Construct', *Ergonomics* 42(11): 1549–60.

Shotter, J. (1983), '"Duality of Structure" and "Intelligibility" in an Ecological Psychology', *Journal for the Theory of Social Behaviour* 13: 19–45.

Vicente, K.J. (1999), *Cognitive Work Analysis: Toward Safe, Productive, and Healthy Computer-based Work* (Mahwah, NJ: Lawrence Erlbaum Associates).

Weick, K.E. et al. (1999), 'Organizing for High Reliability: Processes of Collective Mindfulness', *Research in Organizational Behavior* 21: 81–123.

Weigmann, D.A. and Shappell, S.A. (1997), 'Human Factors Analysis of Postaccident Data: Applying Theoretical Taxonomies of Human Error', *The International Journal of Aviation Psychology* 7(1): 67–81.

Woods, D.D. and Cook, R.I. (1999), 'Perspectives on Human Error: Hindsights, Biases and Local Rationality', in Durso, F.T. (ed.) (1999), *Handbook of Applied Cognition* (Chichester, NY: John Wiley & Sons).

PART I

Identifying System Vulnerabilities within Incident and Accident Analysis

Ger Wackers

Introduction

It is well known that we can learn from incidents and accidents. By studying them we can improve our understanding of the paradoxical processes that produce in work processes both reliable outcomes and system condition drift in which system integrity can no longer be maintained. However, how to learn from accidents or what exactly it is that we can learn from them is not always obvious. Given the challenges facing those who design and those who work within complex technological systems, what alternate explanatory frameworks are available to facilitate learning from incidents and accidents? To what extent does the type of model of the system and of the explanatory framework used by organizations operating complex technological systems, or by accident investigating committees, hamper learning from incidents and accidents? Much depends on the way in which one conceptualizes complex technological systems; that is, on how they are framed. Given that multiple ways of framing and explaining a particular event are possible, multiple ways of framing differ in the mechanisms and processes that they foreground. This occurs because of the theoretical tools and sensitivities that they bring to bear in the analysis.

Part I of this book comprises three chapters that (a) develop alternative methodologies to accident investigations (Norros and Nuutinen, Chapter 2); (b) reinvestigates and reframes an accident in the Norwegian railway system (Rosness, Chapter 3); and (c) accidents related to the North Sea offshore drilling industry (Wackers, Chapter 4).

These studies shift the focus from human error as a (contributing) cause of accidents and emphasize adaptivity in risky work environments; adaptivity that is necessary to achieve reliable performance at an overall system level most of the time, which also, sometimes simultaneously, induces system vulnerabilities.

A Note on Framing

To appreciate how the accounts in these chapters differ from the views and convictions commonly held in operating organizations or accident investigation committees, a note on framing is offered.

Framing is not just the set of concepts or the theoretical perspective with which a scientific discipline (quantitative risk analysis; reliability engineering) approaches its object. It is a recursive process that is reproduced in day-to-day practice. As an outcome of a recursive process the frame shapes interactions and contracts between actors; it shapes experiences, provides cognitive repertoires, suggests courses of actions and provides a conduit for the amplification of consequences – be it innovations or failures – and their movement through the system. Borrowing the concept of framing from Goffman (1974), Callon (1998, 249) argues that:

> [t]he frame establishes a boundary within which interactions take place more or less independently of their surrounding context. ... It presupposes actors who are bringing to bear cognitive resources as well as forms of behaviour and strategies which have been shaped and structured by previous experience.

The framing process, Callon (1998, 249) goes on, 'does not just depend on this commitment by the actors themselves; it is rooted in various physical and organizational devices'. Among these organizational devices is what Callon calls a 'metrological infrastructure' (statistical tools, formats for data extraction and collection, establishment of databases, and so on).

In organizations operating complex technological systems it is common to make a distinction between the technical system and the human workers and social organization, the interface being the domain of human factor analysis. Both technical components and human workers must comply with 'as-designed' functional specifications. Proper maintenance strategies and methods aim to maintain technical system integrity. On the human and social side of the divide much emphasis is placed on the avoidance of human errors and on compliance with rules and regulations. This specific but very common way of framing the system lays the groundwork for asymmetrical explanations of reliable performance, as well as system failure: knowledge, experience and competence going into the engineering and organizational design effort explain reliable performance, whereas deviations from functional design specifications constitute the set of acceptable causes in causal explanations of large-scale accidents. These deviations may be either technical or human/organizational. The reasoning over past decades, is that, for example, work in reliability engineering and tools like reliability-centred maintenance have considerably improved the technical integrity of technical systems: in the large-scale accidents that still occur, human or organizational errors play a significant role.

This way of framing complex technological systems is deeply embedded in operating companies, public regulatory bodies and accident investigation committees alike. It carries with it a confidence and conviction that it is possible to build a reliable organization that is able to operate high-risk technologies in a safe and reliable way. However, in addition to shaping ways of attributing causality and distributing responsibility, this particular way of framing also carries a diminished sensitivity for and understanding of (adaptive) mechanisms and processes that fall outside the frame.

Towards the end of the twentieth century several researchers from a variety of disciplines criticized these ways of framing and associated causal explanatory frameworks. Organizational theorist Charles Perrow proposed a (rather static) framework for the explanation of large-scale accidents in terms of the tight coupling and interactive complexity of these systems (Perrow 1984/1999). For a further discussion of tight coupling, see Rosness (this volume). The three chapters contained in this Part all refer to coupling (and decoupling) and its role as a construct to assist in understanding the various interdependencies involved.

Psychologist and human factor analyst Jens Rasmussen pointed out 'the fallacy of the Defence in Depth Philosophy' that builds on this way of framing and emphasized adaptation and self-organization in work processes, processes that produce drift towards the boundaries of safe operation and ultimately erosion of the precondition for safety (Rasmussen 1990; 1994). A number of detailed case studies of accidents conducted by researchers in the fields of organizational theory and science and technology studies further emphasized not only the importance of understanding processes of adaptation and self-organization, but also highlighted the importance of understanding the historical nature of complex technological systems and the organizations that operate and maintain them (Vaughan 1996; Snook 2000). These ideas about framing are important in assisting us to shift focus beyond traditional error causation approaches. There are four key themes that the contributors to Part I make to this shift:

1. moving away from (human) error causes
2. adaptivity, drift and the emergence of vulnerability
3. developmental history of experience
4. systemic stress and tension.

Moving Away From (Human) Error-causes

The idea that human errors are at least contributing causes to accidents is deeply ingrained in our culture. Several of the following chapters take seriously the fact that often human errors gain social existence in the secondary process of incident and accident investigation. All the chapters pick up and extend the explanatory focus beyond the 'sharp end' of the accident (that is, the workers directly involved). They do so by providing a richer contextual analysis that highlights wider socio-political elements at play in the accident. These include professional identities, transitional organizational change and the effects of global capital.

Adaptivity, Drift and Emergence of Vulnerability

An emphasis on adaptivity and on the emergence of a local standard for good work, and of vulnerabilities is a common theme in Chapter 3 (Rosness); and Chapter 4 (Wackers). The chapters contained in this Part of the book all pick up on the notion of drift and advance its use as an analytic tool.

Developmental History of Experience

The prominence of adaptivity, emergence of local routines and vulnerabilities in several of the following chapters suggests that it is important to think in new ways about history and historical track records. In these chapters the complex technological systems under study are presented as historical entities. Each complex technological system over time traces a unique, historical trajectory, in a commercial and competitive context often driven by strategies to optimize (or sometimes simply maintain an acceptable) performance of the technical system and of the company. These optimizing strategies may induce emergent system vulnerabilities; that is, a reduced ability to anticipate, cope with, resist or recover from 'events' that threaten the achievement or maintenance of functional system integrity.

In his analysis of the Åsta railway accident, Rosness explicitly explores the evolution of vulnerability on the Norwegian Røros railway line. Drawing on various notions and perspectives from organizational theory, he argues that, at the time of the accident, the Røros line was in a transitional state between a traditional manually operated railway line and a fully remotely controlled railway line. More specifically he argues that, over a period of 25 years, the railway company failed to coordinate and integrate decisions made in different parts and divisions of the organization with respect to their consequences for safety on the Røros line. Although each of these decisions may have been defensible – as adaptively appropriate – in the local context and at the time they were made, the gradual decoupling – called 'practical drift' by Snook (2000) – of various decisions in terms of their consequences for rail traffic safety in and through a parallel process of physical and organizational restructuring of the system, induced in the Røros line specific vulnerabilities that were recognized by some but not acted upon by others.

The studies presented here call for a sensitivity to and continuous monitoring of adaptive and emergent processes, and a recognition of the fact that these processes not only induce vulnerabilities, but also help the system to overcome tensions and frictions, for example between divergent goals with respect to punctuality in the provision of services to the public, economic performance of the organization or company, and safety.

Systemic Stress: Tensions and Frictions

Finally, the idea that tension-free systems will never exist and that human work will be necessary to negotiate these tensions in order to produce reliable performance of the system is the last of the common themes that emerges from Part I. Systemic stress may be the result of a kind of inherent incompatibility between diverse system requirements (regularity/punctuality, profitability, safety). It may also be the result of historical changes in the development of identity in work. It may be the result of sudden and profound changes in the environment; for example, a dramatic fall in oil prices inducing an economic gradient favouring less costly field development or design solutions. Or it may be the result of long but only partly coordinated physical and organizational restructuring processes, leaving the system in a perhaps continuous state of transition.

Norros and Nuutinen (Chapter 2) have placed frictions and tension at the centre of their approach. Positioning themselves explicitly in an activity theory perspective, Norros and Nuutinen put the task that has to be performed by, in their case, ship pilots, in the centre of their analysis that also accounts for the changing environment in which that task has to be performed. In what they call Core Task Analysis, the core task comprises the functional demands that are to be fulfilled by the operator for the safe and qualified performance of the system. Due to internal and external pressures for change in these dynamic, heterogeneous socio-technical systems, the core tasks also tend to change, and established, actual working practices that fulfilled the core task at one time may, over time, become invalid. The dynamic development of the system as a whole induces tensions and frictions within the system that are of consequence for the constraints and possibilities the organization provides, or should provide, for safe and efficient piloting. Tension-free systems will never exist, but through systematic analysis and assessment the tensions can be made visible and attempts can be made to reduce them. Norros and Nuutinen draw on Cultural-Historical Activity Theory, which posits that systems may reach a temporary equilibrium until more systemic contradictions (which are always present) come to the fore thus creating developmental opportunities for organizational or institutional development. In their analyses, they provide one possible tool to monitor the adaptive development history of systems, the emergence of new tensions and frictions and the ways in which these are being negotiated by human work.

All the chapters in this section advance many of the current challenges of studying human work within fallible systems. They provide elaborate analyses that reveal the ways in which broader contexts are intertwined with work practice. They elaborate our understanding of important contstructs such as vulnerability and drift and they illustrate why cause-effect models of human error at the sharp end are deficient.

References

Callon, M. (1998), 'An Essay on Framing and Overflowing', in Callon, M. (ed.) (1998), *The Law of the Markets* (Oxford: Blackwell).

Goffman, E. (1974), *Frame Analysis: An Essay on the Organization of Experience* (Cambridge, MA: Harvard University Press).

Perrow, C. (1984/1999), *Normal Accidents: Living with High-Risk Technologies. With a New Afterword and a Postscript on the Y2K Problem* (Princeton, NJ: Princeton University Press).

Rasmussen, J. (1990), 'Learning from Experience? How? Some Research Issues in Industrial Risk Management', in Leplat, J. and de Terssac, G. (eds.) (1990), *Les Facteurs Humains de la fiabilite dans les Systemes Complexes* (Toulouse: Octares Entreprise).

Rasmussen, J. (1994), 'Risk Management, Adaptation, and Design for Safety', in Brehmer, B. and Sahlin, N.E. (eds.) (1994), *Future Risks and Risk Management* (Dordrecht: Kluwer).

Snook, S.A. (2000), *Friendly Fire: The Accidental Shootdown of U.S. Black Hawks over Northern Iraq* (Princeton, NJ: Princeton University Press).

Vaughan, D. (1996), *The Challenger Launch Decision: Risky Technology, Culture and Deviance at NASA* (Chicago: University of Chicago Press).

Chapter 2
Learning from Accidents: Analysis of Normal Practices

Lenna Norros and Maaria Nuutinen

In risky work all opportunities to learn and improve safety should be utilized. For the development of work it is important to know what is occurring on the shop floor – what kind of situational demands exist, what are work practices like for workers, how do development efforts take effect, and so on. Accidents and incidents can be an important information source for answering all these questions. Investigations are accomplished with the intention that both organizational and individual learning from the accidents will take place. To the extent that one can infer learning from changes made as a consequence of post-accident inquiry, however, learning seems to be less effective. On average only 33 per cent of recommendations lead to change according to *Helsingin Sanomat*, the leading Finnish newspaper (26 August 2002). Recommendations are claimed to be either too generic or too detailed. Conducting accident investigations in such a way that contributes not only to explaining the accident but also to enhancing learning within those working in the system or having an impact on its development and thus contributing to the development of the system is a difficult task. There are several recognized challenges for accident investigation (Reason 1998, 52; see also Perrow 1984; Botting and Johnson 1998; Johnson and Botting 1999). Perhaps one of the most difficult problems is the retrospective nature of investigation, and the danger of hindsight. This challenge and the overall difficulty in attaining the view of those who were inside the unfolding situation are recognized by many authors (for example, Dekker 2002; Hollnagel 2002; Woods 1994).

A further challenge in learning from accidents is that accidents are always negative occurrences often accompanied by big headlines in tabloids, showers of accusations, defensive communication, for example pressures for prosecution and liabilities to compensate often complicate the situation. When an accident occurs it is evident that something has gone wrong in the functioning of the particular socio-technical system. In accidents the negative result of particular actions, decisions made, deficiencies of training, for example, become visible. In the case of a negative result or a deviation from an ideal it is possible to reflect on the course of actions and reason how it caused the end result. Learning can take place, but attention can be easily focused on the negative – despite that fact that there could have been even greater losses.

Understanding risky work studies on normal daily work could have advantages over studies on accidents. However, even when analysing normal daily work we have to be able to draw a line between those practices that enhance safety and those that do not in order to steer the systemic development. In normal work situations this task can be even harder than in accident cases because the result does not directly inform whether practices are enhancing or decreasing safety.

The difficulties in learning and in development of risky work are not dependent on the source of data – either accidents or normal situations. If anything, it depends on how we look at the data, the framework and its ability to support reflection. Studies on accidents can inform us about normal practices and can highlight both the vulnerabilities and strengths of the system.

This chapter presents a new method for accident analysis based on a framework we developed originally for reflection of practices in analyses on normal daily work. By keeping the principles we considered important in analyses of daily work, we are able to bring important new elements to accident investigation. Before describing the new method, we focus on two related themes affecting the opportunities for learning: the accident models behind the investigation methods, and analysis of practice in accident investigation.

This accident analysis method was mainly developed while analysing marine accidents. A marine transport system is an example of a complex technological system. Some of its characteristics, however, differentiate it from systems such as nuclear power plants or aviation. Perrow (1984) even calls a marine system an 'error-inducing' system (for example, the attribution of 'human error' is more frequent in this system than in many others) and lists several characteristics of the system that makes it vulnerable: the industry is very old; there are strong traditions; and there are very complex rules and regulations, which are dependent on both international and national interests. Social organization of the personnel aboard ship – the captain's role of centralized control over a small number of personnel; and the problems that arise if the system suddenly expands to include another ship – are quite unique characteristics. There are many economic pressures, and the structure of the industry and insurance is complicated in a marine transport system. All of these characteristics make the marine system very difficult to develop as a whole. Single attempts to improve some parts of it can actually increase the problems of the whole system (for example, increasing of tight-couplings or complexity; see Perrow 1984.) For an explanation of the role of coupling (and decoupling) in systems see Chapter 3.

The Conceptual Basis for this Accident Analysis

Our study took place in the context of an established accident investigation practice. The Accident Investigation Board Finland (AIB) is located within the Ministry of Justice and investigates all major accidents regardless of their nature, as well as all aviation, maritime and rail accidents and their incidents. The AIB defines the purpose of the investigation of accidents as improvement of safety and to prevent future accidents.

The investigations of the AIB have a broad scope. They consider the course of events of the accident, its causes and consequences as well as the relevant rescue measures. Particular attention is paid to the fulfilment of safety requirements in the planning, design, manufacture, construction and use of the equipment and structures involved in the accident. Also investigated is whether supervision and inspection has been carried out in an appropriate manner. Any detected shortcomings in safety rules and regulations may call for investigation as well. In addition to the direct causes of an accident, the accident investigation intends to reveal any contributory factors and background circumstances that may be found in the organization, the directions, the code of practice or the work methods (AIB Annual Report 2006).

The AIB also helps to develop accident investigation methods in collaboration with various international organizations. The ultimate aim of gaining methodological excellence is to improve the learning from accidents and, hence, to effectively prevent new accidents from occurring. The AIB especially emphasizes the use of well-controlled and systematic ways of working in the investigation in order to improve the control of the investigation process and to facilitate the coherence of reasoning.

Without doubt, well-controlled work and usage of systematic methods are important determinants of the quality of investigations. Systematic working is a necessary but probably not a sufficient condition to improve learning from accidents. It is also necessary to improve the conceptual validity of the methods. This was one of the reasons why VTT human factors experts were invited to participate in the investigations. The methodological demands became particularly explicit because, unlike the usual practice in which the analysis work focuses on just one case, the aim in this particular marine accident investigation was to study several accidents in one comprehensive investigation process. While participating in the AIB practice we also identified that good interaction within the interdisciplinary investigation team is an important determinant of the quality of investigation (Nuutinen and Norros 2007). However, the conceptual explicitness and pertinence of the method, upon which this chapter concentrates, is one of the factors that also improve interaction within the investigation team.

Drawing on international research literature concerning organizational and human factors of high risk domains and analyses of accidental events, at least two conceptual issues are especially relevant. The first relates to the accident model used, and the second, intertwined issue, deals with the model of the human actor.

Accident Models

Several authors have drawn attention to the relevance of the adopted accident model for the quality of the investigation and its potential to contribute to the development of the complex socio-technical system (Hollnagel 2002; Leveson 2004). The accident models are a set of assumptions of what comprises the underlying 'mechanisms' are (Hollnagel 2002). Hollnagel provides an overview of the development of accident analyses and the accident models used. He distinguishes three types of models: sequential, epidemiological and systemic. Rasmussen and Svedung (2000) have also proposed corresponding classifications.

The sequential models focus on the causal chain of events that lead to the accident. This way of thinking clearly corresponds to the demands set on the accident investigation to clarify what has happened. The inherent motive connected to the model is to remove the cause.

Epidemiological models are more sophisticated in the sense that they consider the causal structure of an accident more comprehensively. The central concept of epidemiological models is contributing factors (Reason 1990). This relates to the idea that the causal chain may be extended in time and space and that various factors may have latent effects. Constructing defences is the main means to prevent accidents. Defences are various physical, technical, organizational or person-related measures that are designed into the system to prevent accidents. Accidents happen when several such measures, organized in an in-depth defence structure, fail at the same time.

Systemic models are different from the first two model types in that they do not focus on the chain of events but more broadly on the purposeful functioning of the whole activity domain. Behind the systemic models there is the belief that there will always be variability in the system and that the best option therefore is to monitor the system's performance so that uncontrollable variability can be caught (Hollnagel 2002, 1–3). The focus is therefore on the generic regularities according to which singular situational courses of events are constructed under the constraints set by the domain. Only some particular courses of events lead to unwanted results, and most of the time the system functions and people act adequately and even adaptively. Hence, the main concept in this approach is adaptation. This refers to the ability to take account of the significance of situational features with regard to the intended end results, and organize action accordingly. As this type of accident model does not focus on the end result of the activity (incident or accident) that provides a reference from which the investigator could trace back to causes of deviations, we have to seek other bases of evaluation. There is a need to identify what is good action or adaptation in some other way.

Generalizing from Accidents as a Pre-requisite for Learning

The restrictions of the sequential and epidemiological models become more and more recognized when analysing accidents in complex socio-technical systems, and especially in comprehending the human role within them. Other difficulties are, for example, the arbitrariness and theoretical weaknesses of the human error construct (Hollnagel 2002; Lipshitz 1997; Rasmussen 1996; Woods 1994), subjectivity in selecting the events and conditions for analysis (Leveson 2004) and the trap of hindsight (for alternative approaches to accident analysis that also avoid the hindsight trap, see also Chapters 3 and 4). Moreover, investigators are often under pressure to define measures to solve the problems. These may turn out to be local 'fixes' that eventually play a major role in causing the next accident (Reason 1998, 52; see also Perrow 1984). Salo and Svensson (2003) have shown in their analysis of the quality of reporting nuclear power plant events in international reporting systems that, in practical connections, explanations of events typically reduce to very few causal factors. Finally, Nancy Leveson (2004) lists generic transformations in the systems and their contexts, which tend to stretch the limits of methods that are based on event-sequence models. Such transformations include the fast pace of technological change, the changing nature of accidents, new types of hazards, decreasing tolerance for a single accident, increasing complexity and coupling, increasing complexity of relationships between humans and automation, and changing regulatory and public views on safety.

Beyond the above-mentioned problems that the increasing diversity and complexity of the systems put on the existing accident models there is another underlying and more fundamental problem. This concerns sequential and epidemiological models that may be considered closely related because both are based on causal explanations. Hence, both types of models have the tendency to fall into circular reasoning in complex contexts (Rasmussen and Svedung 2000). Moreover, they have a weakness in generalization, because they use concepts that refer to the particular situation (Rasmussen 1996) and not to understanding the constraints and reasons behind events (Leveson 2004).

Systemic models are expected to improve generalization from accident investigation. In systemic models human performance variability is seen as necessary for a user to learn and for the system to develop (Hollnagel 2002). Hence, it is important to distinguish which adaptation is beneficial and which is not (Gauthereau 2003; Hollnagel 2002). This issue is tackled by the newly introduced approach to understanding high-risk organizations called the 'resilience engineering' approach. It is an attempt to develop a conceptual framework to understand risks and develop safety within complex systems (Hollnagel, Woods and Leveson 2006). While attempting to identify the more generic phenomena of high reliability, it amounts to a systemic accident model (Hollnagel 2006). How does the resilience engineering approach improve the generalization of accident analysis results?

The Relevance of the Resilience Engineering Concept in Accident Analysis

Different definitions of the concept of resilience are in use. However, central to all is that resilience denotes an 'ability of a system or organization to keep or recover quickly to a stable state, allowing it to continue operations during and after a major mishap or in the presence of continuous significant stresses' (Wreathall

2006, 275). Woods adds a specification to the concept by noting that resilience is the ability to not only adapt to anticipated fluctuations of the system but also to consider changes and expansions of the system (Woods 2006, 22–23). Inherent for the the effectiveness of the concept is that it supports focusing on the regulating principles that shape the functioning of the system and make it more or less brittle. We summarize the content of resilience engineering in three major points:

1. Model the aims and purposes of the activity to define safe action.
2. Define the safety envelope.
3. Develop an understanding of the specifics of safe action.

First, in Jens Rasmussen's (1997) original claim that appears central to 'resilience thinking', he argued that causal modelling of accidents is insufficient for developing safety because it focuses on a particular course of events. It would, instead, be necessary to develop modelling techniques that focus on the aims and purposes of the activity and identify the result-critical functions and means for their fulfilment. Because such a functional structure can be identified on different levels of abstraction, a hierarchical model may be built of the domain. Such a modelling is used to define the envelope of safe action, which is used as a generic reference in the evaluation of the activity of organizations. Such a modelling approach has been characterized as predictive, or formative (Vicente 1999), because it facilitates identification of safety threats before accidents or incidents become actualized. Hollnagel later confirmed that functional modelling constitutes a predictive aspect of resilience engineering (2006, 17). Also, Leveson (2004, 250) makes use of the functional modelling approach, tackling safety in terms of a control problem and introducing a new accident model based on systems theory. In her model, systems are viewed as interrelated components that are kept in a state of dynamic equilibrium by feedback loops of information and control.

A second characteristic of the resilience engineering approach is that good organizational behaviour is defined as remaining within the limits of the functionally defined envelope of safe actions. The 'drift to danger' model developed by Rasmussen (Rasmussen and Svedung 2000) is an illustration of different pressures and factors that may cause a migration out of the safe area. Descriptions of organizational features that indicate an organization's position with regard to the safety envelope can be found in different publications of the high reliability organization school (HRO), the central figure of which is Karl Weick (1995). Dekker admits that using the 'drift into failure' metaphor to characterize organizational behaviour is intuitively appealing (2006, 85). He makes the point, however, that the hard part is to define how the boundaries are determined. In this connection he proposes two markers of organizational behaviour that could be used to define the borders: keeping discussion about risks, and responses to failure, alive in the organization (Dekker 2006, 75–92). Also Reiman and Oedewald have used the 'safety envelope' and 'drift into failure' metaphors in their outline of an organizational psychological theory on safety-critical organizations (Reiman and Oedewald 2007).

The third point – noted but not yet sufficiently developed within the resilience engineering approach – is grounding the global level analysis of the boundaries of safe action to understanding specific features of sharp-end actions. Hence, Wreathall points out that there is a need to find out how organizations mobilize themselves in critical situations (Wreathall 2006, 278). He also mentions that analysis should shed light on what actually is boundary-crossing activity in a specific situation. A further problem is to identify regularities or maybe modes of daily actions that would explain how capable or motivated people are to take into account the situational constraints. As Klemola and Norros have shown (1997), expert practices differ with regard to focus on situation-specific features. Cook and Nemeth, who accomplished ethnographic analyses in safety-critical work, draw attention to the need to analyze resilience on the level of personal action and decision-making in real situations (2006). The authors claim that at present, the only strong evidence of resilience that they have identified is the presence of certain performances that they call resilience performances (2006, 216). One of the features of resilient performance is the actors' traversing the goal-means hierarchy when facing problems in work (2006, 218–19), for example, by connecting observed specific phenomena to more global functions or checking whether generic safety functions are fulfilled. Both operations indicate that information is integrated into its systemic relationships and in the process made meaningful.

How to Analyze Resilient Performance in Real Situations

For the systemic accident models in general and the resilience engineering approach in particular to be practically helpful for design- or development-oriented aims like accident investigation, there must be a connection between the generic definition of boundaries of safe action and manifestation of safe acts in particular situations (see also Chapters 3 and 4). The method should also allow analysis of how people consider the generic constraints and possibilities and make inferences of their meaning to safety in particular situations. This would require that the individual-level performance could be described in terms of its potential for promoting safety described in the functional analysis.

In order to meet the above-mentioned requirement, we should replace 'action' as the unit of analysis with that of 'practice'. The main reason to change the concept is that action refers to a chain of particular actual performance elements which are defined either in content-independent ways or refer to specific situations. Neither way is sufficient in explaining the process of constructing the course and inherent logic of action in a situation: in the former case the context is neglected and in the latter the specific context is too dominant an explanatory factor. What is needed is identification of a generative and meaningful principle according to which the situation is taken into account and behaviour organized. The concept of practice is a good candidate for this. It denotes the way cognitive resources available in a distributed cognitive system, including the environment and the human actors, are organized. Practice reveals the learned process and the *ways* of doing things (*how*) rather than just stating *what* is done in a particular situation. Hence, practice is a construct that has a predictive or formative nature. This also means that practices can be identified independent of the outcome of a particular course of action. This implies that both normal, successful and accidental courses of action may be analyzed from the point of view of practice. Practices may be identified as good, risky, and so on, independent of the end result in a particular course of events under scrutiny.

The bodily, environmentally, socially and historically distributed cognitive system, which also includes technological elements, organizes itself according to the results to be achieved and according to environmental resources. 'Practice' is the theoretical construct which describes the structuring of this diversely distributed system. The concept of practice has been used in phenomenological and ethnomethodological approaches, not least due to the influence of the seminal book by Lucy Suchman (1987). We make use of this tradition which emphasizes that in the analysis of activity we should take the reasons people give for their actions as the starting point of analysis. We extend the phenomenological notion of practice by drawing on the American pragmatist concept of habit (Norros 2004). According to Peirce (1903/1998), habit portrays a three-part meaning structure comprising sign, object and interpretant. This structure may be used as a basis for identifying habits in empirical analyses as has been shown by Norros (2004). A particular benefit of using the pragmatist habit concept in empirical analysis of practices is that this concept enables the connection of practical operations to their meaning.

Adoption of the notion of practice as a tool to describe how the distributed cognitive resources are organized assumes an even deeper change in our thinking of human–environment interaction and the role of technology. The concept of practice draws on philosophical traditions which aim to overcome the distinction between mind and body or human and environment. Phenomenology, Marxist philosophy and American pragmatism share this basic intention. This is why these traditions are readily referred to when the attempt is to question the relevance of the information processing paradigm that explains human cognition and action as serial goal-driven transformation processes between the two entities, human and environment. We share this intention and see that rather than viewing 'human' and 'environment' as two separate systems, these two should be considered as forming a functional unity that finds its structure according to the targeted results (Järvilehto 2000; see further Norros 2004). This notion has been been advocated in cognitive psychology and studies of human action for example by Hutchins (1995), Ingold (2000) and Dourish (2001). The benefit of the notion of a united, functional human–environment system is that mental phenomena can be seen as connecting certain parts of the environment so that, together with human elements, they form distributed capabilities. Tools and instruments are seen to change the borders of the human body.

The concept of the 'joint cognitive system' proposed by Hollnagel and Woods (Hollnagel and Woods 2006; Woods and Hollnagel 2006) appears to be closely related to the above-described way of comprehending human activity in complex environments. These authors do conceive a co-agency of human and technology and mention that the two are functionally not separate systems (Woods and Hollnagel 2006, 19). However, to be precise, the system Hollnagel and Woods consider is actually not the human–environment system but the human–technology relationship, and the change in perspective is probably not as significant as the change suggested here. As a consequence, Hollnagel and Woods do not question the concept of action explicitly nor do they define the concept of practice as an alternative way of tackling human behaviour in an environment. The authors are interested to define global regularities of the joint cognitive system behaviour that they call 'patterns'. It remains somewhat unspecified how the proposed main types of patterns (resilience, collaboration and affordance) are to be defined empirically, and what defines the features of the human and technical elements that make up the system. However, it is possible to interpret that the concept of pattern is used for the same aims that we have in mind when utilizing the concept of habit.

We have chosen to use the concept of practice and understand practices as habits that are capable of constituting an adaptive process. If one draws on an everyday usage of the notion, adaptiveness does not appear to be a salient feature of habit. Yet, according to the pragmatist theory, habits via their way of functioning do provide a way of acting adaptively in a changing and contingent environment. This so-called abductive way of functioning of habit comprises a continuous cycling between confirming and questioning when acting in the environment which enables making sense of the environment. This continuous, and not necessarily a consciously controlled, process is an interpretative one. It focuses on the specific and singular features of the situation. While doing so, general and confirmed beliefs are made use of, and new beliefs are created. In this way, habits enable survival in a contingent world. In pragmatist terms, resilient performance is interpretative acting, which is behaviour that is organized to make full use of the abductive potential of habits. This potential is central for learning from singular experience.

Making use of the above-described theoretical underpinnings we claim that interpretiveness is the first generic evaluation basis of practices. Beyond this, it should be necessary to make use of a more context-related evaluation basis, the core task. We analyze whether situated acting is focused on the functionally significant demands of the domain when fulfilling the aims of the activity. This evaluation basis has been labelled as core-task orientedness (Norros 2004; Savioja and Norros 2007).

Identifying Practices Involved in Accident Situations

A systemic accident analysis method was developed that portrays the principles highlighted above. It is capable of considering the actual online performance of the actors involved in the accidents. A systemic approach requires that the analysis is capable of considering the interaction between context-dependent and person-related factors. Hence, it is necessary that the constraints and possibilities of the domain and the particular situation are first considered from a holistic perspective. Referring to the different accident models, a systemic accident model should be adopted. This is more or less a top-down point of view. In order to accomplish an analysis of local activities and actual decision-making of actors in a system-oriented way, it is necessary to find out what aims and purposes the actors were considering and which constraints and opportunities they took into account while acting and taking decisions. To fulfil these aims we made use of a two-way systemic approach for the analysis of human activity in dynamic, complex and uncertain environments, called the Core Task Analysis (CTA) (Norros 2004).

The core task concept and analysis methods have been developed in process control studies accomplished at VTT Technical Research Centre of Finland (Norros and Nuutinen 2005; Klemola and Norros 1997; Nuutinen 2006). It has been extended by organizational culture studies (Reiman and Norros 2002; Oedewald and Reiman 2003; Reiman and Oedewald 2004, Reiman and Oedwald 2007). The CTA framework is grounded in the cultural-historical theory of activity (Engeström 1999; Leontjev 1978; Vygotsky 1978) but it also draws on American pragmatism and its conceptions of human conduct and comprehension in a continuously changing environment (Peirce 1903/1998, Lecture IV (179–97) and Lecture VII (226–41); Dewey 1999). The framework has adopted further methodological principles from the ethnomethodologically-oriented approaches of situated action (Suchman 1987; Hutchins 1995) as well as from research on work domain modelling (Vicente 1999) and from the analysis of safety-critical domains (Hollnagel 2002; 2004). (For more detail see Norros 2004.)

The core task is defined as 'the shared objectives and the outcome-critical content of work that should be taken into account by the actors in their tasks performance for maintaining appropriate interaction with the environment' (Norros 2004, 17). The environment refers to the physical, occupational, social, political and economic environment. By referring to the objectives and outcome-critical content the core-task concept creates a connection between the *activity of* an organization and the *activities in* an organization. It is the conceptual method used to describe objectives and hoped-for outcomes, intrinsic constraints of work and psychological work demands, and how these could be, and effectively are, fulfilled by the operators (or organizations) in situated actions for the efficient performance of the system.

Core Task Analysis methodology aims at revealing significant behavioural regularities to identify these in both successful and unsuccessful courses of action. Analysis is not focused on the end result of actions, but rather it is employed to make inferences from the particular course of action that produced it. In this regard, the aim is not to declare *what* happened as such but to uncover *how* people used their resources and for what reasons, or *why*.

This chapter describes how this generic CTA approach was applied to accident analysis. We first used the CTA approach in connection with analysing several air-traffic control incidents (Norros and Nuutinen 1999; AIB B 8/1997; AIB 2/1993). Its second use was in the context of an extensive investigation of maritime accidents in which we participated. In the latter phase Core Task Model-based indicators were used in a more systematic and extended way than in the first phase (Norros, Nuutinen and Larjo 2006; Nuutinen and Norros 2007) and the Expert Identity construct (presented in detail in Nuutinen 2005a as part of the model; see also Nuutinen 2005b; 2006). In this connection we also exploited the activity system model in the analysis of the organizational background of the piloting activity.

Why Piloting Accidents are so Frequent

Several foreign ships were involved in accidents during piloting in the Finnish fairways in the autumn of 1997. These accidents were interpreted as indicative of underlying problems in the piloting activity in the restricted coastal waters of Finland. Therefore, AIB decided to launch a joint investigation of ten accidents that had taken place between 1997 and 2000 with the aim of creating a more comprehensive understanding of any common reasons behind the accidents. Each accident case was investigated and the results were published in the primary accident investigation reports (AIB C 11/1997 M; AIB C 15/1997 M; AIB C 16/1997 M; AIB C 1/1998 M; AIB C 4/1998 M; AIB C 5/1998 M; AIB C 13/1998 M; AIB C 9/1998 M; AIB C 11/1998 M; AIB C 2/2000 M[1]). In the second phase of investigation the results regarding each accident were compared and synthesized, results and conclusions being published in a separate report (AIB S 1/2004 M).

Analysis of Accidents: Main Steps and Results

In our applied investigations, the aim is to analyze particular accident cases from the perspective of what they tell us about current practices in the domain under investigation. In particular, we aim to find out in which direction the evolution of the practice should be steered and by what means. The path of the investigation and the questions to be answered are presented in Figure 2.1 and the developed method is then described in more detail. The present method consists of six phases linked to each other. In the following the content of each phase is explained. (The phases of examination are also explained in Nuutinen and Norros 2007.)

---

1   C 11/1997 M, C 4/1998 M and C 5/1998 M also available in English (http://www.onnettomuustutkinta.fi/2601.htm).

Steps

1. Tracing the 'facts' of the accident:
- The what, where and when of each accident?

2. Interpreting and modelling the core task of piloting:
- What the accidents tell about the objectives and constraints of piloting and its psychological work demands – that is, the core task of piloting?

3. Assessing the resources and situational demands:
- What are the constraints of and resources for piloting in comparison with the core task?

4. Assessing human performance during the accident voyage:
- How well the core task was taken care of with the available resources?
- What are the current practices?

5. Understanding piloting practice from an activity system perspective:
- What the accidents and other material tell about the evolution of the practices?
- What the accidents and other material tell about the development tensions in the piloting system?

6. Formulating conclusions and recommendations:
- What kind of recommendation promotes the development of the piloting system as a whole?

Result: For further analysis /
For an accident report

What?
Course of events
Core-task model:
General indicators of good performance
Course of events of the piloting trip

How?
Situational criteria for good performance
Comparable descriptions of resources and practice /
Conditions for navigation and piloting
Navigation method
Cooperation on the bridge

Why?
Tensions in activity system
Conclusions and recommendations

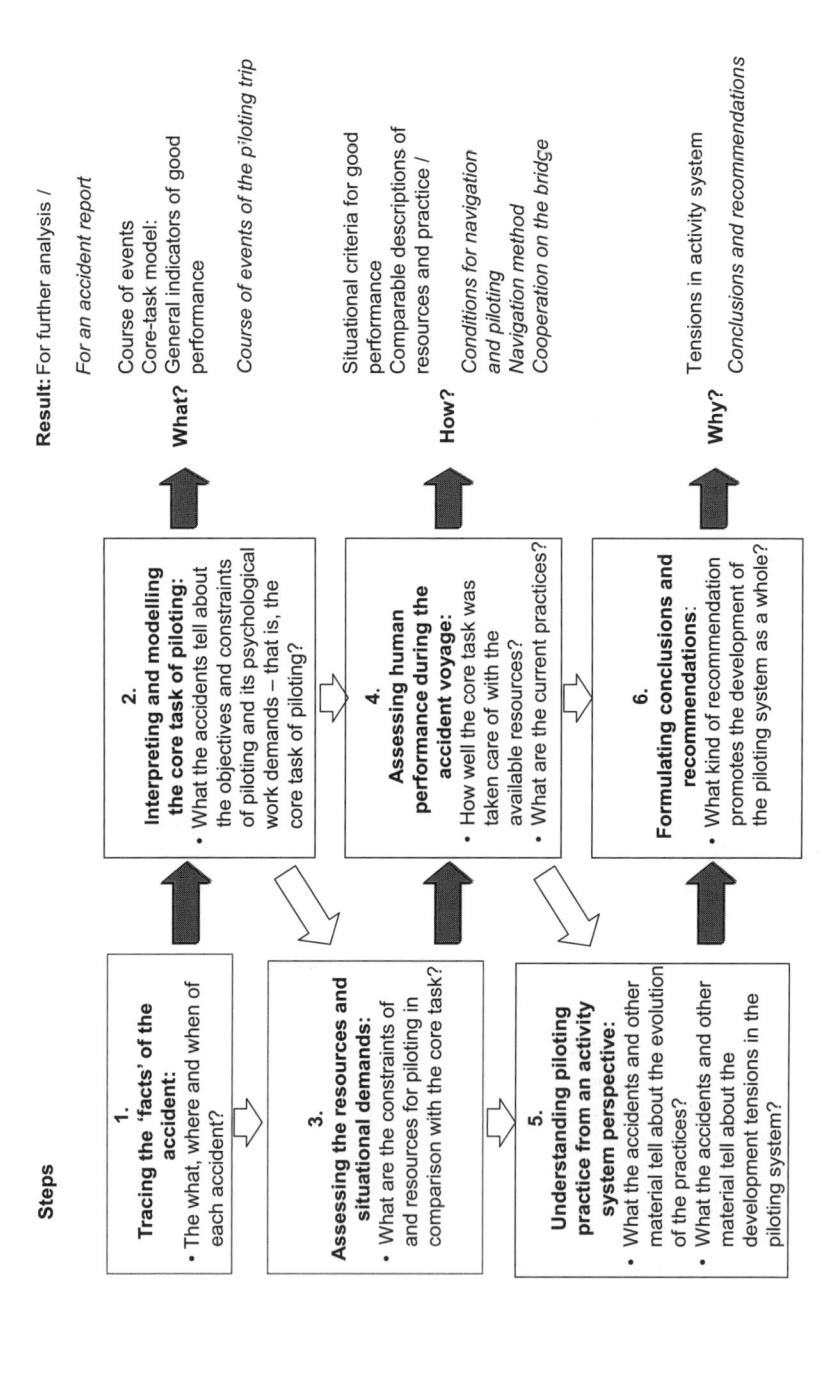

Figure 2.1 The path of the accident investigation and results of the phases

Tracing the Facts of the Accident

The first task in the investigation is to find out the 'what', 'where' and 'when' of what has happened. Each single accident is first described in the light of the acquirable evidence. The aim is not to achieve 'the true nature of the situation' in order to find what people had missed (Dekker 2002), but to describe the course of events and their context. Based on the premises of the Core Task Analysis framework, the context of the accident was first analyzed as accurately as possible.

The first phase includes the description of the vessel, the crew, the traffic limitations, the cockpit and its equipment, the weather conditions and the sea area. The course of accident events, rescue activity (not included in further analyses here) and damage to the vessel are also described. The participants' actions are described in their temporal order. The intention is that this information could be considered 'facts' and that they should be described in detail and remain objective, despite their suggested contribution to the accident. An important challenge is to keep the interpretation open as far as possible in order to avoid the tendency to seek facts that strengthen the hypotheses.

Modelling the Core Task of Piloting

When reflecting on the current piloting practices and deciding the direction in which they should be developed, it is necessary to consider the task of piloting in a new way. The piloting task needs to be considered from a wider, more comprehensive perspective than that enabled by traditional task analysis. What are the aims of piloting activity and what are the intrinsic constraints and possibilities of the work domain that set more or less contradictory demands on how people may act? The second phase aims to accomplish such an analysis of piloting and to define good performance in piloting on a general level (see Table 2.1).

The modelling was based on available material concerning the aims, constraints and opportunities of piloting in the Finnish fairways. The conceptions of actual practitioners should be carefully taken into account and new ways must be created to involve the practitioners in the definition of the core task. This issue is stressed also by McCarthy et al., who warn of overweighting the explicit organizational goals and of neglecting the goals and concerns of those working in the organization (McCarthy et al. 2004). Previous studies and writings on sea pilotage were used as a starting point. The field study on pilotage that was accomplished earlier at VTT provided an important source for the modelling. In this previous work a first core-task modelling was carried out on the basis of expert interviews, and observations on 15 normal piloting situations were accomplished in situ (Norros et al. 1998). Otherwise there were only very few studies on sea pilotage that provided appropriate support for the modelling task. The relevance of these studies to the Core Task Analysis was limited because they usually focused on a single factor or issue. Moreover, many of them (such as analysing the effects of alcohol are carried

Table 2.1 The core task model of sea piloting

AIM: efficiency	Critical Functions	Interactions	Constraints and Possibilities	Working Practice Indicators
PRODUCTIVE / SAFE / PROMOTES WELLBEING	Control of complexity: vessel diversity, several actors, part of the traffic system / Coping with uncertainty: hydrodynamic uncertainty and uniqueness of ship-sea system / Control of dynamics of moving system and delay in steering	**Operator(s) process** *Adoption to situational demands*	Characteristics of water area / Control: delayed / Control: mediated, unique and complex / Information: uncertain, distributed / Information: direct and mediated, different representations	1. Creating knowledge of a particular sea area 2. Creating a situational orientation 3. Anticipation and waiting for results 4. Testing ship's controllability 5. Maintaining orientation in the moving vessel 6. Controlling the movements of a vessel 7. Taking account of the whole traffic situation 8. Information formation on a cumulative interpretation of the situation and location 9. Integrating information from different representations 10. Shifting from one representation to another
		Operator – operator *Transparency Common norms and procedures* *Common knowledge and experience base*	Many tasks, distribution required but varied / Actors from different organizations / Specialized knowledge	11. Formation of and updating shared plans 12. Formation of shared interpretation of the situation 13. Following norms and common practices 14. Exercising control over performance through monitoring and re-checking 15. Integrating expertise regarding the environment, routes and the vessel
		Facilitating cooperation and building trust	Changing, new cooperative partners	16. Constructing and maintaining an ad-hoc team 17. Taking account of the work demands in timing of changes in division of work
		To own actions *Meaningfulness closely related to core task*	Core task changes, not easy to define / Organizational change and new pressures (e.g. service work)	18. Continuous attempts reflect on one's own conception of the core task 19. Focusing on the core task in demanding situations

SITUATIONAL, PHASED (preparation, team construction and piloting voyage) CRITERIA

Table 2.1   Concluded

AIM: efficiency	Critical Functions	Interactions	Constraints and Possibilities	Working Practice Indicators	
PRODUCTIVE / SAFE / PROMOTES WELLBEING	Control of complexity: vessel diversity, several actors, part of the traffic system / Coping with uncertainty: Hydrodynamic uncertainty and uniqueness of ship-sea system / Control of dynamics of moving system & celay in steering	*Realistic self-confidence*	Breaking up old community: weakening peer support for evaluation New cooperation partner every time Responsibility vagueness between bridge crew and pilot Expertise based on specialized knowledge and experience	20. Continuous attempt reflect on one's own conception of the core task 21. Focusing on the core task in demanding situations 22. Balancing between contradicting goals 23. Courage to make decisions in demanding situations 24. Constructing an understanding of a bridge crew's competence 25. Balance between 'proving' own expertise and submitting it to monitoring 26. Reflectivity, attempt to learn and analyse own competence and its restrictions	SITUATIONAL, PHASED (preparation, team construction and piloting voyage) CRITERIA
		Sensibility for sense of control (in order to adapt one's behaviour)	Feedback on actions: delayed, deficient (two-way: mediated: equipment/direct: environment)	27. Balance between acting according to plan – situation 28. Seeking different sources for feedback 29. Preparedness for possible difficulties and high demands	

Source: Adapted from Nuutinen and Norros 2007

out in training simulators. A further problem is that studies are not generalizable, since the conditions for piloting are quite different in different areas around the world. The Finnish coast is long and characterized by narrow fairways and rocky waters, and piloting in the archipelago area is unique.

Every accident under investigation was analyzed with regard to what it could reveal about the critical functions of the controlled process (ship–sea system), and the characteristics of the operational interactions between the operator and the process, between the operators (cooperation) and the relationship to a person's own actions and self. When the various interactions of a piloting performance were interpreted against the aims and functions that create meaning to their accomplishment, a number of psychological work demands could be inferred. Fulfilling these demands constitutes the contextual evaluation bases, the core-task orientedness, for good practice.

As indicated in our earlier discussion on practice, behind the operational interactions in fulfilling the core task there are some general assumptions about what kind of features characterize 'good piloting' in an uncertain, complex and dynamic environment. The key assumption is that in order to survive in such an environment, people create beliefs and repetitive forms of action, the relevance of which is tested in action against situationally varying conditions (Peirce 1903/1998, Lecture IV (179–97) and Lecture VII (226–41); Ingold 2000; Hutchins 1995 and Norros 2004). As a result of an embodied interpretative work, a generic understanding of the environment is achieved. Adaptation and innovation draws on such ongoing interpretative effort. In the operator–process interaction, the adaptability of actions to situational demands is pertinent for controlling the hydrodynamic uncertainty of the sea–ship system in order to ensure the system's efficiency.

As a result of the two evaluation bases, 'core-task orientedness' and 'interpretativeness', general criteria were inferred. These are applied with regard to three forms of interaction that is required when dealing with complex, dynamic and uncertain environments. These are operator–process, operator–operator and operator–self interactions. Because the expectation was that pilotage must demand collaboration and teamwork, a lot of attention was devoted to defining the operator–operator interactions. Several general criteria to analyze this interaction were identified (Hutchins 1995; Norros et al. 1998; Nuutinen and Norros 2001). The first general criterion is transparency of performance (see Table 2.1). This refers to the ability of the bridge crew and the pilot to perceive what is going on. At least three aspects have an effect on transparency of performance: transparency of the goal, interaction and tools. The second general criterion is sharing common knowledge and experience base. A bridge crew and a pilot should have at least a partially shared common knowledge and experience base in order to be able to interpret each other's behaviour and to adapt their own behaviour. The third general criterion comprises common norms and procedures and a commitment to them (Table 2.1). There is a need for norms and regulations in situations with a great deal of uncertainty and several independent actors. Every traffic system provides an example of the importance of norms. The fourth tentative general criterion is facilitating cooperation and building trust (see Table 2.1). This criterion is based on the study reported herein but has also emerged in a recent study of expert services in industry (Nuutinen 2004; 2005b). The cooperative partners in piloting are usually strangers to each other, in comparison with, for example, crews at power plants. They are also members of different organizations and thus do not share a common training or common practices. Cooperation in such situations calls for extra social effort to make it work. The emphasis on building trust does not mean that the demands of ensuring activity by double-checking and monitoring should be neglected.

Behind the third interaction there is the expert identity construct (suggested by Nuutinen 2005a; 2006), which aims to capture the emotional–energetic aspects intertwined with both work performance and the development of expertise at work (Norros and Nuutinen 2002). The first general criterion of this interaction is that a person's feeling towards the meaningfulness of his or her work is closely related to the aspect of the core task relevance. The second criterion is realistic self-confidence, which, in practice, means that a person has a quite realistic opinion of his or her competence and is ready to ask for help when needed. The third criterion is the sensibility for the sense of control. Sense of control refers to the situational control of one's actions and the emotions awakened by reaching goals (through exercising control). As Table 2.1 indicates, the operator–process interaction could in this case be captured only by one generic indicator, adaptation to situational demands.

The core task of piloting emerged and 27 indicators were defined to portray its various aspects (see Table 2.1). In this investigation several accidents were used as source material in the core-task modelling. The idea, however, is that the core task should be also included in a normal singular accident investigation. All available material should always be used in order to reach an understanding of the core task.

The outcome of phases 1 and 2 are twofold (see Figure 2.1). First, the chain of events can be made explicit. Hence, detailed descriptions of the courses of events and actions in each particular accident can be acquired. Such a description is a causal, outside observer's view of events. The descriptions were published in the primary accident investigation reports referred to above. Second, a general core task model of the piloting activity emerged. The aims of the activity and the constraints and opportunities that set boundaries for tasks needed to fulfil the intended end results were defined in this model (see Figure 2.1).

Assessing the Resources and Situational Demands in the Accident Cases

Assessment of resources and situational demands refers to the functional specification of the constraints for navigation and piloting in a particular archipelago fairway with a particular ship. This includes assessment of investigation material, which can reveal a rich body of knowledge. Drawing on it, the core-task demands are mapped to the actual condition of the pilotages which were described according to the three main phases of the piloting performance: the preparation phase, the team-construction phase and the voyage phase.

This phase of the investigation focuses on the specific features of each accident. Thus, an assessment of the organization of piloting as it is accomplished in each piloting district that was involved in the accidents should be accomplished. The practices and procedures of the shipping company involved in the piloting situation should also be clarified. The analysis also includes considering the fairway geometry in comparison with the particular ship's characteristics (size) and the navigation marks, weather conditions, bridge layout, steering and navigation equipment, and available number and quality of the bridge crew. The fairway design is very important because it sets the constraints for accomplishing a turn. The possibilities and constraints for making a safe turn in the fairway were tested with ship manoeuvring-simulation software running on a PC. The simulator tests were used in order to define intervals at the point of turning, the required elevator angle, the minimum speed at which the vessel can still manoeuvre and the effects of wind on the ship's drift. In addition, the possible contradictions between statements about timing in the different investigation material were tested with the simulator whenever possible.

Navigation equipment and the information content of such equipment were analyzed from a sequential task point of view: what possibilities and constraints do the artefacts offer for taking care of the sequential task and fulfilling the core-task demands? Here the definitions of the tasks – steering, controlling engine, navigating, and so on – (Norros et al. 1998) were used as an assessment base in addition to the core-task model. This analysis was related to other situational conditions, for example manning the bridge and the possibilities of distributing tasks.

The outcome of this phase of the investigation is the estimation of the situational conditions for efficient piloting. The results achieved indicate that there are problems in the conditions for piloting (see Table 2.2). First, several problems in the fairways were recognized. For example, even on the new fairways along the Finnish coast there are turns whose specifications deviate from international recommendations. A further important issue is how the implementation of the new fairways is organized. The markings of the sea fairways are not always consistent, which may lead to false interpretations of available safety margins. There are also many problems with the suitability and usability of the navigation equipment with respect to piloting. Demands set by navigation on the open sea seem to dominate the development of the equipment while the special demands of navigation in the archipelago fairways are mostly neglected. However, some new equipment, for example electronic mapping, that is being implemented on some of the ships and becoming standard equipment could create good resources for piloting, but only if it is updated with the archipelago fairways and actually used. The analyzed accidents raised the issue that many ships do not have an adequate manual steering system, which would allow both adaptive and controlled turns.

In summary, the conditions were adversely affected in three of the ten cases. There were also cases where the conditions were good for safe piloting and the difficulty of the situation itself cannot explain the accidents (see Table 2.2).

Assessing Behaviour in an Accident from a Practice Point of View

Analysis and assessment of the behaviour of people involved in the accident is conducted in this phase. The first aim of the analysis is to find features of the current practices and ways of acting that may be defined as risky. Such ways of acting might have previously served well but could in the present conditions be found inadequate or even harmful. The second aim is to find features that help to compensate weaknesses in the initial conditions of safe piloting. The resulting characterizations should identify generic features of current piloting practice and

Table 2.2 **Issues that complicated the conditions for safe piloting in the different cases**

Cases. Reduced resources or increased demands because of:	1	**2**	3	4	**5**	6	7	**8**	9	*10*	Total %
The design and/or maintenance of fairways	X	X	X	X	X		X	X			70
Interface or quality of radar or GPS		X		X	X						30
Inadequate or problematic steering equipment	X	X	X		X	X	X	X			70
Demanding weather conditions	X	X	X	X	X			X		X	70
Manning on the bridge		X						X			20
	3	5	3	3	4	*1*	2	4	*0*	*1*	

GPS = Global positioning system

The cases with clearly reduced conditions for safe piloting are marked in bold. Italics refer to the cases with none or only one issue reducing the conditions for safe piloting.

Source: Adapted from Nuutinen and Norros 2007

the aim is to test whether these could explain the specific behaviour of the actors in each accident.

The method used to identify generic tendencies that could explain actual realizations is to explore whether and according to which logic the core-task demands are taken care of. As a reference we exploited the Core Task Model and the 27 indicators presented in Table 2.1.

The possibility of making comparisons between and generalizations from single cases is based on the systematic use of the functional core-task oriented indicators and classifying the evaluations according to a specified scale. When there is evidence that the demand has been taken into account, even when it needs extra effort (for example, because of poor resources), the behaviour is valued as 'excellent' (2) on this indicator. The degree 'adequate' (1) is given when there is evidence that the demand has been taken care of and there is no evidence to suggest otherwise. The assessment 'poor' (0) is prescribed when it is evident that the demand was actual but was not taken care of. When the data is limited, and no evidence of the matter of concern is found, the indicator is, of course, not used. Simple statistical operations and graphical illustration were used in order to summarize the results of the different indicators and the assessment of cases in the investigation. The summaries of the cases were also compared with the summaries of the resources and the demands in order to raise the level of analysis.

Results concerning piloting practices

The analysis of the piloting practices was accomplished mainly by collaborative evaluation work by three members of the investigation team using the above-mentioned rating measure. The detailed results including ratings with regard to each indicator are presented in Nuutinen and Norros (2007). In this context we restrict ourselves to a more global view of the results, as described below.

The ten accident cases were divided into three groups according to the mean value of all the indicators. Cases included in group A (1, 2 and 4) demonstrated the best level of practice (over 40 per cent of the indicators had a value of 1 or 2); cases included in group B (7, 8, 9 and 10) were an intermediate group; and group C (3, 5 and 6) demonstrated the worst level of practice (under 20 per cent of the indicators had a value of 1 or 2 and the percentage of value 0 was highest). Thus, in group A the pilots' performance was quite good according to most of the indicators (see Figure 2.2). In group B some of the demands were managed in an excellent manner and others quite poorly. There were also many missing values (empty cells) in group B. In group C the performance was at a quite poor level according to most of the indicators. When considering the results of the three groups in the three interactions, pilot–process, pilot–captain and pilot–self, it was found that the third – the expert identity-related indicators – suffered most in groups B and C.

It is clear that differences in the performance of the bridge crews could also be a result of different situational conditions (and not differences in working practice). When compared with the conditions for safe piloting (see Table 2.2 and Figure 2.2), there was no systematic dependency between the difficulty of the conditions and ranking of the cases. In conclusion, the differences in performance could be interpreted as denoting different piloting practices.

How historic forms of navigation explain differences in piloting practices

In order to explain the results concerning the piloting practices and their connections to the variation of the conditions of safe piloting, we explored the history of navigation. The starting point was Edwin Hutchin's study (1995) in which he explored navigation activity from the perspective of anthropology and cognitive psychology. Hutchins

Learning from Accidents 37

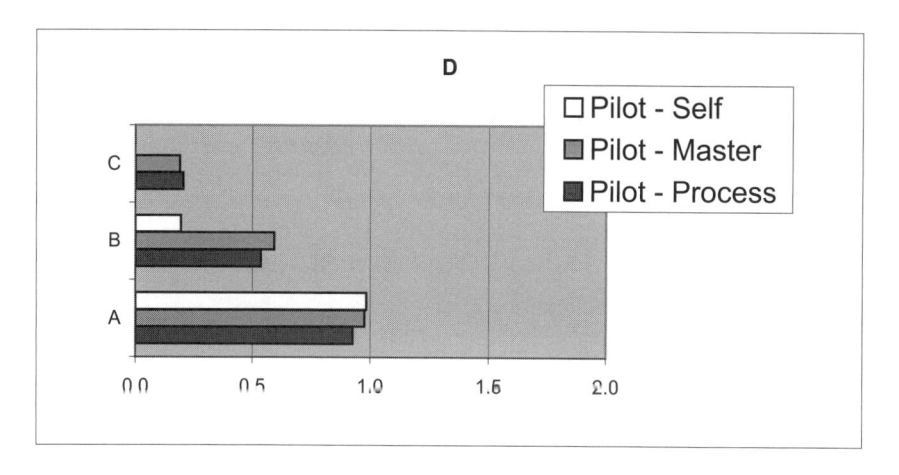

Figure 2.2 The pilots' performance three groups (A–C)

contrasted the old Micronesian way of navigating with that typical of Western countries in open-sea navigation. The Micronesian navigator keeps all information he needs in his mind, whereas the Western navigator uses physical navigation tools. Drawing on these studies the investigation board abstracted two types of practice: the traditional and the technologically mediated way of navigating (AIB S 1/2004 M). An 'inside-out' or actor-centred perspective on the environment is typical of the traditional way of navigating. Moreover, the task is represented sequentially using immediate visual information on stable references, and memorized and otherwise implicit tools and knowledge. A navigator's individual performance has a central role. The navigators take care of all of the tasks by themselves in this way of navigating. In contrast, a 'bird's-eye' perspective on the environment is typical in the technically mediated way of navigating. This is realized through geometrical representation of the task using technical artefacts and external tools and concepts. The external representation of the task in artefacts opens up possibilities for shared awareness of the situation and collaboration among the bridge personnel. These two different ways of navigating were used as a frame of reference when aiming at understanding the differences in the ways of piloting as described by our indicators, and the non-monotonous interaction between the performance level and the difficulty of the conditions. We proposed that current piloting practice is dominated by features typical of the traditional way of navigating. Weaknesses in this way of navigating under the present demands could explain the noted problems and the accidents. The current piloting practice shares many of the features of traditional navigation (see Figure 2.3):

- The navigation task is conceived and described as a serial one-dimensional process;
- Immediate visual perceptions from the surrounding environment of the ship are compared with each other;

- The frame of reference is tacit and understanding of the route is based on experience and maps that the pilots have explicitly memorized;
- The pilot's individual performance has a central role; they perform every task by themselves and do not see a need to share them or to communicate their intentions or knowledge, or to control (ensure) actions.

The strength of the traditional way of navigating and the corresponding piloting practice lies in the situational adaptability of the action it facilitates. Its weakness is its tacit nature and the fact that it does not facilitate collaboration.

Features of piloting practice that resemble those of a technically mediated navigation are (see Figure 2.3):

- There is a bird's-eye view of the environment;
- The task is described by geometrical concepts;
- The task performance utilizes measuring and computational operations;
- Navigation equipment mediates the relationship to the environment and also records experience from the environment and the route;
- The potential of navigation technology is utilized in cooperation and distributed cognition to enable monitoring, re-checking and division of tasks, and this is not regarded as a threat to expert identity.

An explanation of how piloting practices interact with the conditions of safe piloting is depicted in Figure 2.4. The poor performance of group C can be explained by weaknesses in the traditional way of navigating in demanding situations. In such a situation the use of all of the available (albeit impaired) resources, extra effort in fulfilling the core task and such expert identity that allows efficient cooperation becomes crucial. The tendency seems to be to regress even more to the traditional pilot-centred way when demands increase during piloting. The performance profile of group B indicates that the traditional way of navigating is also quite strong in that group, although in less demanding situations this group may still deliver a reasonably good performance. In addition, when examining the situational criteria closer, the peaks in the performance profile can be further explained by compliance with procedures at the time concerning piloting and known quality criteria of good piloting culture – for example, route plans, showing them to the master, following fairways, using a helmsman when entering a narrow fairway area. It seems that the importance of new demands on navigating practice had been accepted in this group but the very core of practice still remains traditional. Group A tends to combine traditional and new technical features in navigational methods. The performance of those bridge crews included in this group demonstrates more explicit cooperation and new types of expert identity, which, in turn, contributes to better process control. However, in demanding conditions even this emerging, new piloting practice could not compensate for the high demands of the situation and poor resources – and an accident took place.

Learning from Accidents 39

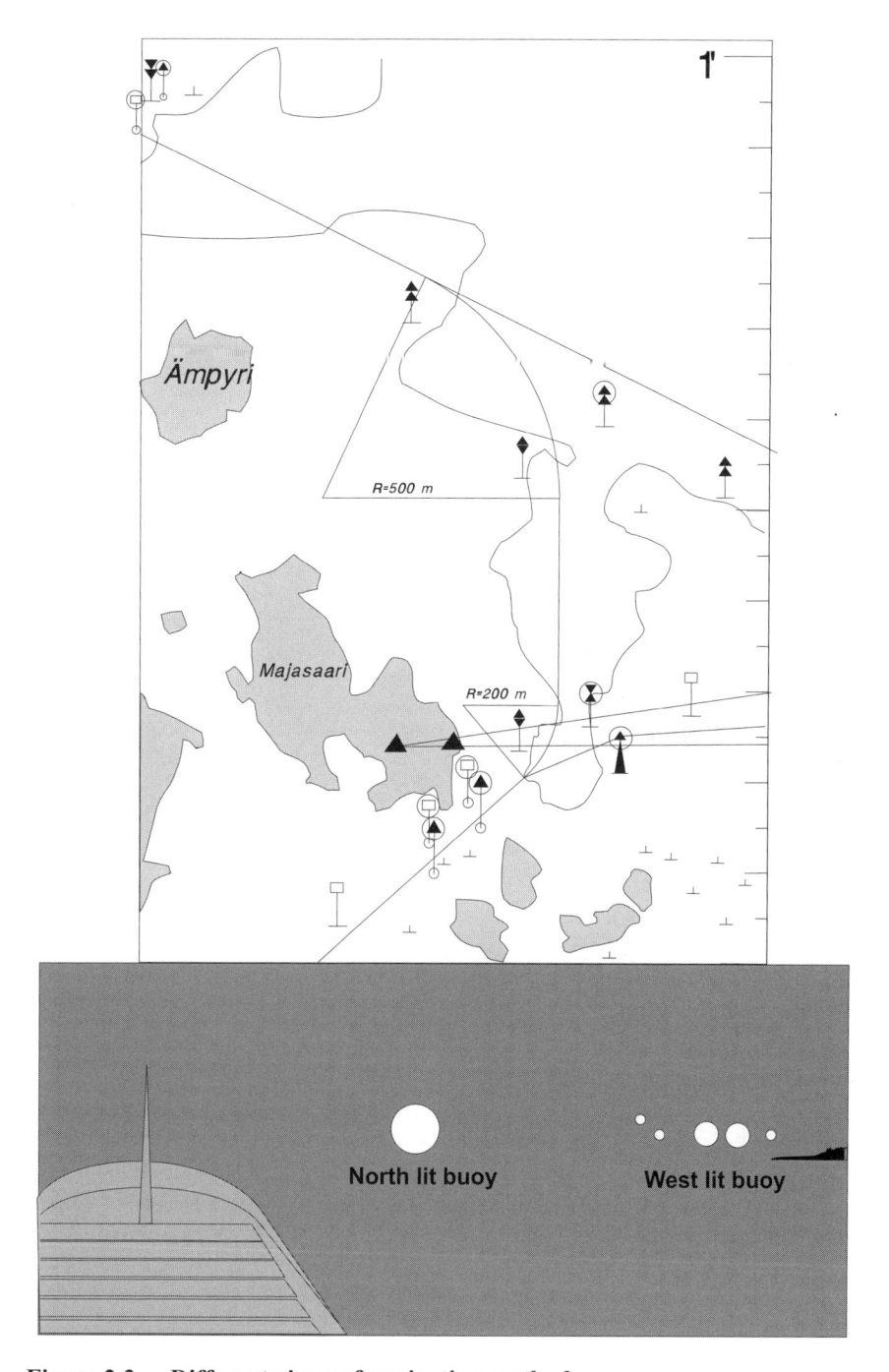

Figure 2.3   Different views of navigation methods

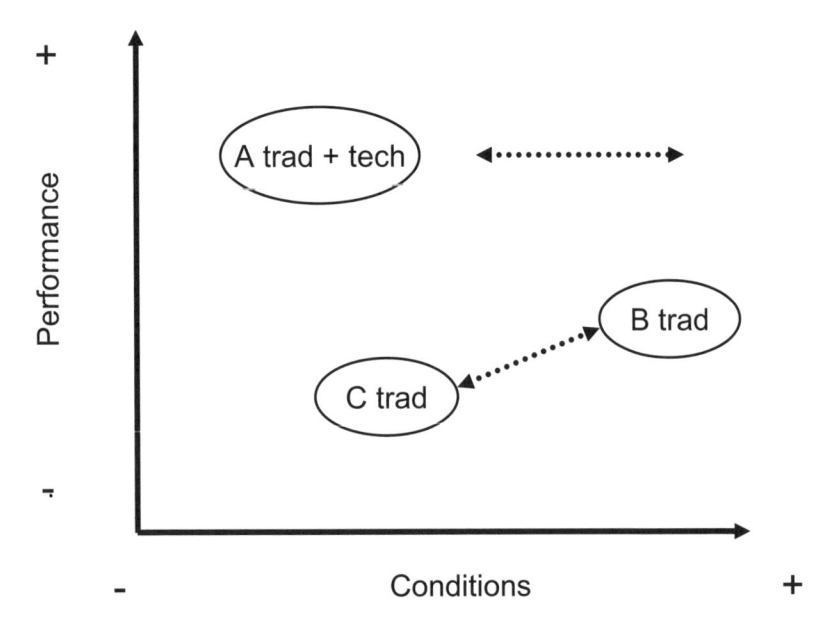

Figure 2.4 Interpretation of the differences in the rated piloting practices (performance) with the aid of historical types of navigation

Understanding Piloting Practice from an Activity System Perspective

In the next phase of the investigation we broadened our scope of analysis to a system level in order to understand why new forms of practice are not adopted. The general targets and constraints of an activity constitute the invisible context of the situated actions and make them meaningful and understandable. Equally, contradictory tendencies that are typical in the evolution of any activity system become evident as tensions and disturbances in the functioning of the system in particular situations (for a different discussion on tension in situated action, see also Chapter 5). Our idea was that a conceptual analysis of piloting as an activity system would probably facilitate understanding of the studied accidents from a broader perspective as ruptures that the tensions of the whole system generate (Engeström 1987; 1999). Figure 2.5 presents an activity system model of the piloting. An activity system consists of the subject (pilot and bridge crew), community (pilots and ship's crew) and object of the activity (ship–sea system). In complex work the activity interactions between these elements are mediated via artefacts, rules and division of labour. As the activity is directed towards the outcome, it greatly affects the organization of the activity.

Historical documents concerning sea piloting, its legislation, meaning in society, functions and organization, and so on, were analyzed. Understanding of the evolution of the practices was gathered by combining the results of the performance assessment with the tensions (see below). The summary investigation

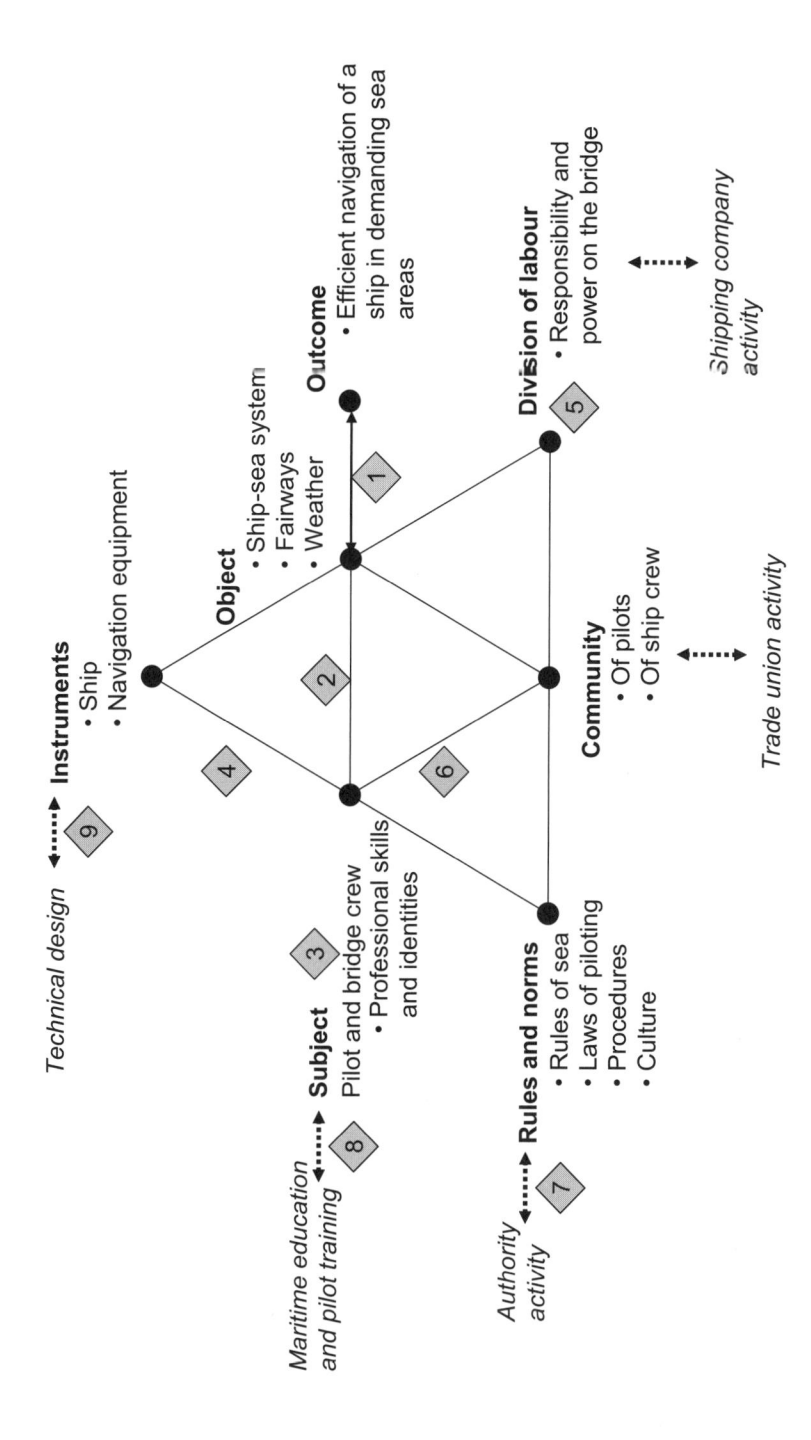

Figure 2.5 Piloting as an activity system

report and its appendices include the evidence used to explain these tensions (AIB S 1/2004 M)

The analysis of piloting as an activity system identified nine tensions (see the numbers in Figure 2.5):

1. The capacity and conditions of the fairways and the traffic system are not able to fulfil the increasing demands for safety, productivity and quality in society in every situation.
2. A reduction of the safety margin is evident whenever bigger ships enter the coastal fairways. Maintaining safety requires technically mediated navigation. This way of navigating includes development of cooperative piloting practices on the bridge. An increase in the interaction with other actors in the traffic system could also be facilitated.
3. The strength of the current piloting practice lies in its skill-based adaptability. The practice does not, however, include communication and cooperation among the bridge crew to the degree that the current demands require.
4. The current piloting practice cannot efficiently take advantage of the existing navigation technology, and the technology is not being developed according to the demands of the piloting task (see point 9).
5. The on-bridge responsibility and power relations (based on professional skills) between the master and the pilot are in contradiction with those enacted in the law. This does not manifest itself until the situation demands it.
6. The sources of the pilots' expert identity seem to be strongly in individual skills and the actors' own experience and responsibility. Relying on these features gets even stronger when the demands of the situation increase. Cooperation is not included in the conception of expertise or the construction of professional identity. The reduction in peer support (resulting from the fact that the time needed at pilot stations has been reduced) can prevent the construction of a new kind of identity when, for example, the observations on the changes in the core task are not shared.
7. Authority control and supervision do not create sufficient support for anticipating difficult situations. Decision-making takes place at the bridge and has been left too much the individual pilots' responsibility.
8. The training of pilots cannot efficiently fulfil the pressures for development. The problems are deficient theoretical teaching in connection with practical training and the lack of basic education on route planning and use of technologies.
9. The demands of piloting are not taken into account when designing the ships' navigation technology. The demands of piloting should be part of the validation and verification criteria of bridge technology.

While not having an opportunity to intrude very deeply into the history of Finnish piloting, the investigation group formulated the hypothesis that the difficulties in cooperation in the existing practices of piloting have historical origins. Throughout Finnish history, especially in times of war, pilots have had an important role in guarding Finnish waters. They have maintained and protected the knowledge on sailable routes. Piloting has become a highly individual and tacit practice, the art of which has only been mediated within piloting families in earlier times, and even up to very recent times piloting has largely been learned in an apprenticeship from expert to novice within a normatively defined and restricted occupation.

Conclusions and Recommendations

On the basis of the previous phases, conclusions and recommendations for each case were formulated. As is usual in accident investigations, the conclusions provided an insight into the particular chain of events that led to each accident. The chain of events describes the causal situational construction of the accident, step-by-step, from the outside observer's point of view. The consideration of contributing factors includes conclusions regarding the resources and the demands for piloting. The conclusions regarding the navigation practice explain the chain of events. The arguments for the recommendations concerning practices and other issues gain support from the results of the activity system analysis.

This and the former phase (5) together create explanations of *why* the accident(s) took place: both in a particular and in a generalized sense. The cases portray tensions within the activity system. The connecting link between the particular and the general are the practices of navigation that portray the culture and habits of the whole navigation community. As a result of our analysis, problems in individual cases can be put into systemic and historical perspective.

Core Task Analysis Approach in Supporting Learning from Accidents

Vulnerabilities in the piloting system were found in the analysis guided by the new method of accident investigation. The results of the investigation indicated that the available conditions for piloting are partly deficient. Current practices have clear strengths but also many weaknesses. Drawing on a historical typology of navigation practices we could conceptualize the differences in practices we had identified empirically with our new rating method of practices. We were also able to formulate predictions concerning the possible brittleness of practices when the conditions of piloting become more demanding. We offered a system-level analysis of pilotage and revealed a number of tensions that may explain why the development of new practices is hindered. These tensions also create pressure on the system and may eventually facilitate changes in many of the practices within the system to better meet the developing needs of safe and efficient navigation. Our analyses allowed us to formulate a hypothesis that piloting practice with strong traditional features is probably more sensitive to situational demands, and performance may tend to deteriorate if conditions worsen. We found evidence for the assumption that a combination of traditional and technically mediated practices would also maintain a good performance level when conditions deteriorate.

Increasing the Resilience of Normal Practices

With reference to the theory of adaptive behaviour in safety-critical systems (Hollnagel et al. 2006) we may note that the currently emerging new piloting practice that could be articulated here would have the function of improving resilience within the system. Using the terminology of Cook and Nemeth (2006) this practice could provide resilient performance.

There is increasing concern about the effects of tightening productivity demands and the importance of monitoring, predicting and controlling an optimization process (Rasmussen 1997), or the impacts of Efficiency-Thoroughness Trade-Off (Hollnagel 2002). The importance of recognizing which adaptation is beneficial and which is not is emphasized in the current discussion of coping with risky environments (Gauthereau 2003). Resilience of performance is an attempt to define what good adaptation could be. We claim that the core-task modelling methodology offers a new solution to define the features of good practice. It is evident that these features are different depending on the domain but that generic features may also be assumed.

The habit-based analysis that we developed for the evaluation of practices is incorporated in the core-task analysis approach. This means that the evaluation of practices refers to system-level boundaries of successful action. When based on the modelling of system boundaries the evaluation is also contextual. The first evaluation dimension for good practice that we have proposed is core-task orientedness. This dimension fulfils both attributes, systemicity and contextuality. In the core-task analysis approach we exploit the concept of habit. Habit is proposed by the American pragmatists to be a developing scheme of action that develops in a practical confrontation with the world. The appropriateness of habits in ensuring continuity in action while coping with the contingencies of particular situations demonstrates an internal characteristic of habit, its interpretative strength. Interpretativeness is the vehicle for adaptive behaviour. Hence, a second, more practice-internal and content-independent evaluation dimension of practices is their level of interpretativeness.

Our evaluation of practices that is founded on the above-mentioned two dimensions does not focus on the outcomes of actions (the situational realizations of habits) but on the content-dependent and internal features of acting appropriately and fulfilling the ethos of the profession (MacIntyre 1984). Such criteria are also informative in cases where the outcome does not indicate insufficient task performance.

The core-task oriented analysis of practices is a way of distinguishing what is beneficial adaptation and what is not. In the analysis of the goodness of practices these are also put into historical connections. The method supports finding those historic features of the current practices that are worth maintaining and reinforcing, and those which could be risky. The expert identity construct used tried to reach the emotional–energetic basis of practices, which does not change rapidly and the significance of which can be the most relevant when meeting high demands and difficulties during piloting. In addition, we tried to find a promising trajectory for future development, since the socio-technical system to be controlled is dynamic – as noted by Leveson (2004).

Generalization as a Challenge for Accident Analysis and Learning from Accidents

In his famous book *Normal Accidents*, Perrow (1984) criticized accident investigators for their inability to take account of economic and other realities of the domain system when making recommendations. Our method provides one possibility for comprehending the particular behaviour in a more comprehensive context, as indicated in the previous section. Perrow's criticism, however, denotes another problem as well: the difficulty of drawing generic conclusion on the basis of singular accident analyses. Rasmussen (1996) also drew attention to this issue when considering different accident models. This issue was tackled and the aim was to develop a method that would improve possibilities for generalization. Our solution was to utilize the habit-based concept of practice, and to define practices using the systemic concept. Both serve to improve generalization according to the methodological solution that we adopted. First, we accomplished the modelling of the domain via a formative modelling method that is focused to identify the generic constraints of the domain. The constraints define the envelope of appropriate action. The tensions we found in our analysis are processes that tend to shift the borders of this envelope.

Practices demonstrate a formative approach to behaviour. By denoting what aspects of the environment are significant for the actors and for which reasons, they deliver knowledge of behavioural inclination and peoples' readiness to act in the work situations faced. The style (mode or way of acting) reveals a tendency to behave in the studied environment and this style, for example practice, is learned in the community of practice in which people act. Due to the formative nature of the habit-based analysis of behaviour, our method enables interpretations of the specific events of accident cases from a generic perspective. For the same reason, this type of analysis also allows the prediction of the appropriateness of a certain type of performance and its vulnerability to particular situational constraints.

The presented method outlined here consisted of six phases. The structuring phase of the investigation is the definition of what could be considered good piloting – that is, the core task of piloting. On this basis we assessed both the situational conditions and the participants' performance in a way that allowed comparisons and generalizations from a single accident, as illustrated in the results. The reasons behind the observed actions could be analyzed and assessed with reference to the core-task demands and their interpretation. In order to explain the unexciting interaction between the core-task-related ways of piloting and the situational conditions we drew on a distinction between two basic ways of navigating. We could reach a credible generic explanation for the phenomena observed in the accident cases. Furthermore, we were able to recognize tensions in the activity system of piloting. We demonstrated that many system-level problems affect the actions in actual piloting situations because the actors are bound to strike a balance between the conflicting aims and demands the global problems create. The solutions become overt in that the practices of the actors indicate what is considered a meaningful basis for action. In order to make sense of complex situations, these practices are repeated.

A significant practical limitation of the analysis of the piloting activity was the scarcity of investigation material concerning the actual courses of action and persons' conceptions of their work. In particular, the analysis suffered from a lack of material on the actions and opinions of the masters or other members of the bridge crew. This restricted the use of the indicators in the assessment. More details of the results of this investigation have been presented to various interest groups. The results have been accepted with interest and development activities have been planned and started. Thus, the results may be claimed to have pragmatic validity, credibility and plausibility (Hammersley 1990).

Learning from Accidents

Concrete indicators and criteria for the evaluation of practices can be defined by using this new method. The evaluation draws on *what* people have done in actual situations, but the more important feature of the method is that by connecting the actual action to the possible and efficient system-related reasons for these actions (*why*), it is possible to reveal *how* people act. Being confronted with the task of defining the reasons and distinguishing differences in ways of acting, people are provided with the opportunity to reflect on their action. As has been argued by Clot (forthcoming), drawing attention to how one works opens up a real learning process. The type of indicators and criteria we have developed may be used both in the development of competencies and in the design of technical and organizational solutions. As the methodology used in accident analyses may also be applied in the analysis of the normal flow of action, we acquire a diversity of compatible results, which may be used for learning and development purposes. In high-risk organizations, where accidents are infrequent but the potential damage to people and the environment is considerable, the possibility of reflecting on normal daily work to derive safe practices is clearly preferred.

References

AIB 2/1993. ATC incident at Helsinki-Vantaa airport, Finland, 29 October 1993.
AIB B 8/1997 L. ATC incidents near Vihti VOR Radio Beacon, Finland, 25 October 1997 and 20 August 1997.
AIB C 11/1997 M. Ms GRIMM, Grounding outside port of Kotka, 1 October 1997.
AIB C 15/1997 M. Ms MARIE LEHMANN, Grounding on the fairway to Tammisaari, 21 November 1997.
AIB C 16/1997 M. CRYSTAL AMETHYST, Grounding off Mussalo harbour in Kotka, 1 December 1997.
AIB C 1/1998 M. Pusher barge MEGA/MOTTI, Grounding at Alrlsto, 5 January 1998.
AIB C 4/1998 M. Ms. GERDA, Grounding outside port of Kotka, 7 April 1998.
AIB C 5/1998 M. Ms BALTIC MERCHANT, Grounding in Puumala at Hätinvirta, 21 April 1998.
AIB C 13/1998 M. Ms TRENDEN, Grounding off Rauma, 17 December 1998.
AIB C 9/1998 M. Ms CHRISTA, Grounding off Kotka, 23 November 1998.
AIB C 11/1998 M. Ms GARDWIND, Grounding off Kotka, 5 December 1998.
AIB C 2/2000 M. Ms AURORA, Dangerous incident and grounding south of Helsinki pilot station Harmaja, 6 March 2000.
AIB S 1/2004 M (2006), Piloting practices and culture in the light of accidents. English summary.
AIB Annual report (2006), Helsinki.
Botting, R.M. and Johnson C.W. (1998), 'A Formal and Structured Approach to the Use of Task Analysis in Accident Modelling', *International Journal of Human-Computer Studies* 49(3): 223–44.
Clot, Y. (forthcoming), 'Clinic of Activity: The Dialogue as Instrument', in Sannino, A., Daniels, H. and Gutierrez, K. (eds.), *Learning and Expanding with Activity Theory* (New York: Cambridge University Press).
Cook, R.I. and Nemeth, C. (2006), 'Taking Things in One's Stride: Cognitive Features of Two Resilient Performances', in Hollnagel E. et al. (eds.) (2006), *Resilience Engineering: Concepts and Precepts* (Aldershot: Ashgate).
Dekker, S.W.A. (2002), 'Reconstructing Human Contributions to Accidents: The New View on Error and Performance', *Journal of Safety Research* 33(3): 371–85.
Dekker, S. (2006), 'Resilience Engineering: Chronicling the Emergence of Confused Consensus', in Hollnagel E. et al. (eds.) (2006), *Resilience Engineering: Concepts and Precepts* (Aldershot: Ashgate).
Dewey, J. (1999), *The Quest for Certainty: A Study of the Relation of Knowledge and Action* (Helsinki: Gaudeamus).
Dourish, P. (2001), *Where the Action is: The Foundations of Embodied Interaction* (Cambridge, MA: MIT Press).

Engeström, Y. (1987), *Learning by Expanding: An Activity-Theoretical Approach to Developmental Research* (Helsinki: Orienta).

Engeström, Y. (1999), 'Activity Theory and Individual and Social Transformation', in Engeström Y., Miettinen, R. and Punamäki, R. (eds.) (1999), *Perspectives on Activity Theory* (Cambridge: Cambridge University Press).

Gauthereau, V. (2003), *Work Practice, Safety and Heedfulness: Studies of Organizational Reliability in Hospitals and Nuclear Power Plants* (Linköping: Unitryck).

Hammersley, M. (1990), *Reading Ethnographic Research: A Critical Guide* (London: Longmans).

Hollnagel, E. (2002), 'Understanding Accidents – From Root Causes to Performance Variability', *IEEE 7th Human Factors Meeting* (Scottsdale, Arizona).

Hollnagel, E. (2004), *Barriers and Accident Prevention* (Aldershot: Ashgate).

Hollnagel, E. (2006), 'Resilience – the Challenge of the Unstable', in Hollnagel, E. et al. (eds.) (2006), *Resilience Engineering: Concepts and Precepts* (Aldershot: Ashgate).

Hollnagel, E. and Woods, D. (2006), *Joint Cognitive Systems: Foundations of Cognitive Systems Engineering* (Boca Raton: Taylor & Francis).

Hollnagel, E., Woods, D. and Leveson, N. (eds.) (2006), *Resilience Engineering: Concepts and Precepts* (Aldershot: Ashgate).

Hutchins, E. (1995), *Cognition in the Wild* (Cambridge, MA: The MIT Press).

Ingold, T. (2000), *The Perception of the Environment: Essays on Livelyhood, Dwelling and Skill* (London: Routledge).

Johnson, C.W. and Botting R.M. (1999), 'Using Reason's Model of Organisational Accidents in Formalising Accident Reports', *Cognition, Technology and Work* 1: 107–118.

Järvilehto, T. (2000), 'The Theory of Organism-Environment System (IV): The Problem of Mental Activity and Consciousness', *Integrative Physiological and Behavioral Science* 35(1): 35–57.

Klemola, U.-M. and Norros, L. (1997), 'Analysis of the Clinical Behaviour of Anaesthetists: Recognition of Uncertainty as a Basis for Practice', *Medical Education* 31: 449–56.

Leontjev, A.N. (1978), *Activity, Consciousness, and Personality* (Englewood Cliffs, NJ: Prentice Hall).

Leveson, N. (2004), 'A New Accident Model for Engineering Safer Systems', *Safety Science* 42(4): 237–70.

Lipshitz, R. (1997), 'Naturalistic Decision Making Perspectives on Decision Error', in Zsambok, K. and Klein, G. (eds.) (1997), *Naturalistic Decision Making* (Mahwah, NJ: Lawrence Erlbaum).

MacIntyre, A. (1984), *After Virtue: Study in Moral Theory* (Notre Dame, IN: University of Notre Dame Press).

McCarthy, J., Wright, P. and Cooke, M. (2004), 'From Information Processing to Dialogical Meaning Making: An Experimental Approach to Cognitive Ergonomics', *Cognition, Technology and Work* 6: 107–116.

Norros, L. (2004), *Acting Under Uncertainty: The Core-Task Analysis in Ecological Study of Work*. (Espoo: VTT Publications No. 546). Available online at: <http://www.vtt.fi/inf/pdf/publications/2004/P546.pdf>

Norros, L. and Nuutinen, M. (1999), 'Development of an Approach to Analysis of Air Traffic Controllers' Working Practices', *Human Error, Safety and System Development. Liege, BE, 7–8 June 1999*.

Norros, L. and Nuutinen, M. (2002), 'The Concept of the Core Task and the Analysis of Working Practices', in Boreham, N., Samurcay, R. and Fischer, M. (eds.) (2002), *Work Process Knowledge* (London: Routledge).

Norros, L. and Nuutinen, M. (2005), 'Performance-Based Usability Evaluation of a Safety Information and Alarm System', *International Journal of Human–Computer Studies* 63(3): 328–61.

Norros, L., Hukki, K., Haapio, A. and Hellevaara, M. (1998), *Decision Making on Bridge in Pilotage Situations*. Research report in Finnish. Summary in English. VTT research report 833, Espoo.

Norros, L., Nuutinen, M. and Larjo, K. (2006), *Luotsauksen Toimintatavat ja Kulttuuri Onnettomuustapausten Valossa* (Helsinki: Accident Investigation Board Finland). (Navigation practices and culture in the light of accidents).

Nuutinen, M. (2004), *Challenges of Remote Expert Services: Modelling Work and Competence Demands at the Case Study*. Research Report in Finnish. Summary in English. VTT research report.

Nuutinen, M. (2005a), 'Expert Identity Construct in Analysing Prerequisites for Expertise Development: A Case Study in Nuclear Power Plant Operators' On-the-Job Training', *Cognition, Technology and Work* 7(4): 288–305.

Nuutinen, M. (2005b), 'Contextual Assessment of Working Practices in Changing Work', *International Journal of Industrial Ergonomics* 35(10): 905–930.

Nuutinen, M. (2006), *Expert Identity in Development of Core-Task Oriented Working Practices for Mastering Demanding Situations: Six Empirical Studies on Operator Work* (Helsinki: Department of Psychology, University of Helsinki).

Nuutinen, M. and Norros, L. (2001), 'Co-operation on Bridge in Piloting Situations: Analysis of 13 Accidents on Finnish Fairways', in Onken, R. (ed.), *CSAPC'01. 8th Conference on Cognitive Science Approaches to Process Control. 'The Cognitive Work Process: Automation and Interaction'*, Munich, 24–26 September 2001.

Nuutinen, M. and Norros, L. (2007), 'Core-Task Analysis in Accident Investigation: Analysis of Maritime Accidents in Piloting Situations', *Cognition Technology and Work* 5, <http://www.springerlink.com/content/t522k2w4181m4785/?p=cc35639890fd494895d8244f7bd78b0a&pi=0>

Oedewald, P. and Reiman, T. (2003), 'Core Task Modelling in Cultural Assessment: A Case Study in Nuclear Power Plant Maintenance', *Cognition, Technology and Work* 5: 283–93.

Peirce, C.S. (1903/1998), *The Harvard Lectures on Pragmatism: The Essential Peirce. Selected Philosophical Writings. Volume 2* (Bloomington, IN: Indiana University Press).

Perrow, C. (1984), *Normal Accidents: Living with High-Risk Technologies* (New York: Basic Books).

Rasmussen, J. (1996), 'Risk Management in a Dynamic Society: A Modelling Problem', *Keynote Address: Conference on Human Interaction with Complex Systems*, Dayton Ohio.

Rasmussen, J. (1997), 'Risk Management in a Dynamic Society: A Modelling Problem', *Safety Science* 27(2/3): 183–213.

Rasmussen, J. and Svedung, I. (2000), *Proactive Risk Management in a Dynamic Society* (Karlstad, Sweden: Swedish Rescue Services Agency).

Reason, J. (1990), *Human Error* (Cambridge: Cambridge University Press).

Reason, J. (1998), *Managing the Risk of Organizational Accidents* (Aldershot: Ashgate).

Reiman, T. and Norros, L. (2002), 'Regulatory Culture: Balancing the Different Demands of Regulatory Practice in the Nuclear Industry', in Kirwan, B., Hale, A. R. and Hopkins, A. (eds.) (2002), *Changing Regulation: Controlling Risks in Society* (Oxford: Pergamon).

Reiman, T. and Oedewald, P. (2004), 'Measuring Maintenance Culture and Maintenance Core Task with CULTURE-Questionnaire – A Case Study in the Power Industry', *Safety Science* 42: 859–89.

Reiman, T. and Oedewald, P. (2007), 'Assessment of Complex Sociotechnical Systems – Theoretical Issues Concerning the Use of Organizational Culture and Organizational Core Task Concepts', *Safety Science* 45(7): 745–68.

Salo, I. and Svensson, O. (2003), 'Mental Causal Models of Incidents Communicated in Licensee Event Reports in a Process Industry', *Cognition Technology and Work* 5: 211–17.

Savioja, P. and Norros, L. (2007), 'Systems Usability: Promoting Core-Task Oriented Work Practices', in Law, E., Hvannberg, E.T. and Cockton, G. (eds.) (2007), *Maturing Usability: Quality in Software, Interaction and Value Systems Usability – Promoting Core-Task Oriented Work Practices* (Springer).

Suchman, L.A. (1987), *Plans and Situated Actions: The Problem of Human–Machine Communication* (Cambridge: Cambridge University Press).

Vicente, K.J. (1999), *Cognitive Work Analysis: Toward Safe, Productive, and Healthy Computer-Based Work* (Mahwah, NJ: Lawrence Erlbaum Associates).

Vygotsky, L.S. (1978), *Mind in Society: The Development of Higher Psychological Processes* (Cambridge, MA: Harvard University Press).

Weick, K.E. (1995), *Sensemaking in Organizations* (Thousand Oaks, CA: SAGE Publications).

Woods, D. (1994), *Behind Human Error: Cognitive Systems, Computers, and Hindsight* (Cseriac Soar, 94-01 Army Navy Air Force FAA Nato. Wright-Patterson Air Force Base, Ohio).

Woods, D. (2006), 'Essential Characteristics of Resilience', in Hollnagel, E. et al. (eds.) (2006), *Resilience Engineering: Concepts and Precepts* (Aldershot: Ashgate).

Woods, D. and Hollnagel, E. (2006), *Joint Cognitive Systems: Patterns in Cognitive Systems Engineering* (Boca Raton: Taylor and Francis).

Wreathall, J. (2006), 'Properties of Resilient Organisations: An Initial View', in Hollnagel, E. et al. (eds) (2006), *Resilience Engineering: Concepts and Precepts* (Aldershot: Ashgate).

Chapter 3
Derailed Decisions: The Evolution of Vulnerability on a Norwegian Railway Line

Ragnar Rosness

On 4 January 2000, two passenger trains collided on the single-track line between Rudstad and Rena on the Røros railway in eastern Norway. The engine car of the northbound train was completely wrecked, while the steering car received minor damage and remained upright on the rails. The southbound locomotive train was severely damaged. The locomotive toppled over onto its side and the front carriage buckled and derailed. A major fire broke out immediately in the area around the locomotive and the wreck of the engine car. A few minutes later the fire spread to the front carriage of the southbound train and eventually to the remaining two carriages. Out of a total of 86 persons on board the two trains, 19 persons were killed. Thirty other passengers – none with critical injuries – were taken to hospital.

How could such an accident happen? Answers differ, depending on one's perspectives and interests. The focus in this chapter is not on whether the accident was caused by human or technical errors, but rather to show how the Røros line became vulnerable to human and technical errors as a consequence of management efforts to meet demands on performance and efficiency in an organization with insufficient capacity to follow up and implement its own decisions, and in the absence of strong watchdogs defending safety interests. Viewed from this perspective, the Åsta accident may sensitize us to how problems in decision-making and implementation can weaken a system's defences and create vulnerabilities.

In order to build this argument, we need to examine the defences against train collisions related to the Åsta accident: what defences were in place, and how did some of them fail? The next step is to place the status of defences at the time of the Åsta accident into the context of the technical evolution of the Røros railway. The Røros line was in a state of transition from an old-fashioned railway controlled by train dispatchers on each station, towards a modern main line with remote controlled signals and switches. The Røros line remained for several years in a vulnerable intermediate stage where some of the defences associated with old-fashioned manual traffic control had been removed, whereas not all defences associated with fully equipped modern main lines had been implemented. Problems related to decision-making and implementation may contribute to the emergence of vulnerability. The latter analysis will be performed in two stages. First, I will summarize relevant findings in the report of the public commission, and identify features of the decision processes that may have contributed to the evolution of vulnerability on the Røros line. Second, I will try to reframe these findings in order to examine how normal organizational processes can contribute to the evolution of vulnerability. This second analysis will draw on conceptions of drift in safety science and organizational theory.

The main information source for this chapter is the report of the public commission that investigated the Åsta accident (NOU 2000:30). I have obtained some additional information on the decision processes related to the accident through discussions with persons who worked in the Norwegian State Railways or the National Railway Administration, from the report of the accident commission of the National Railway Administration (Jernbaneverket 2000), and from a book on the recent history of Norwegian railways (Gulowsen and Ryggvik 2004). The discussion in this chapter is limited to conditions related to the events leading to the collision. Issues related to the capacity of rolling stock to withstand collisions or to the notification and rescue operations after the collision are not discussed.

Drift in Sociotechnical Systems

Several authors have conceptualized the idea that sociotechnical systems tend to drift towards an unsafe state if no controls or counter pressures are established. Using Brownian movements of molecules as a metaphor, Rasmussen (1994; 1997) developed the idea that human activities tend to migrate towards the boundary of acceptable risk. This metaphor may be extended to a situation where several actors approach their perceived boundaries of acceptable risk, possibly without realizing how their activities may interact. Reason (1997) introduced a similar metaphor: 'navigating the safety space'. He proposed that organizations tend to drift passively towards increasing vulnerability unless they are actively driven towards the resistant end of the safety space by energetic implementation of effective safety measures.

Hollnagel (2004) coined the term 'Efficiency-Thoroughness Trade-Off' (ETTO) to capture the ways in which people adapt under conditions such as irregular and unpredictable system input, variable demands, inadequate resources, unexpected behaviour of other people and sub-optimal working conditions. Under unfavourable conditions, ETTOs may lead to high output variability and an increasing gap between design assumptions and work as it is in reality. Hollnagel further proposed the metaphor 'stochastic resonance' to capture the idea that output variations in several processes may interact and create a major disturbance or accident.

Perrow (1984) proposed that some sociotechnical systems have structural properties that are conducive to system accidents. High interactive complexity may lead to unexpected event sequences and make it difficult to diagnose abnormal system states. 'Tight coupling' means that a system lacks 'natural' buffers, so that disturbances propagate rapidly throughout the system, and there is little opportunity for containing disturbances through improvisation. Whereas Perrow thought of complexity and coupling as rather stable structural properties of sociotechnical systems, Weick (1990) argued that these attributes tend to change during periods of crisis or high demand. For instance, the collision between two jumbo-jets at Tenerife airport in 1977 happened on a day when the airport was extremely crowded, it had to handle very large aircraft on a narrow runway, and visibility was poor. These factors made the system much more complex and tightly coupled than on an ordinary day. Moreover, several errors occurred in communication between the tower and the two aircraft involved. Weick argued that these errors caused the system to become even more interactive and tightly coupled. A change in the pattern of coupling may be more dangerous than a system that is constantly tightly coupled, because individuals and groups may carry on with interaction patterns that are adapted to a loosely coupled system.

Snook (2000) devised a complex model of 'practical drift' based on the concept of 'practical action'. 'Practical action' refers to locally efficient action acquired through practice and maintained through unremarkable repetition. Practical drift involves two dimensions of change. One dimension concerns the pattern of tight and loose couplings, that is, strong or weak interdependence between sub-units. The other dimension concerns rule- versus task-based logics of action – going by the book versus adapting behaviour to the local practical circumstances. Snook proposed that sociotechnical systems tend to cycle through four states:

1. The 'designed' organization, which follows global rules in a tightly coupled world. This is because sociotechnical systems are designed to tackle critical situations when couplings are tight.
2. The 'engineered' organization, which initially operates as designed. The global rules are followed initially, but most of the time coupling may be looser than the design basis. This state is unstable because the global rules are likely to be perceived as unnecessarily rigid.
3. The 'applied' organization, where local, task-based logics take over in a loosely coupled environment. Action is gradually adapted to the local conditions, and gradually deviate from the global rules.
4. 'Failure', which may occur if a tightly coupled situation occurs when local, task-based action prevails. Although the sociotechnical system may have been designed to handle the tightly coupled situations, it is no longer operated according to the initial global rules, and may thus be vulnerable.

We now possess a considerable range of constructs to explain drift at the sharp end, among the people working close to the hazard. We do not have a similar array of constructs to account for drift at the blunt end of the organization.

Summary of the Event Sequence

On 4 January 2000, a northbound train from Hamar was scheduled to meet a southbound train from Trondheim at Rudstad station on the Røros line. The northbound train entered the station area and stopped at the platform at 13:05, allowing a passenger to enter the driver's cabin. The train had not yet moved sufficiently far into the station area to allow a meeting train to pass. At 13:07, approximately three minutes before its scheduled departure time, and before the southbound train had arrived, the northbound train left Rudstad.

The southbound train was about eight minutes delayed when it left Rena at approximately 13:07. At 13:12:35, the two trains collided at Åsta, midway between Rudstad and Rena. The speed of the northbound train was 90 km/h, and there is no indication that the brakes were activated. The speed recorder of the southbound train was destroyed in the fire, but its speed at the time of impact was estimated at approximately 80 km/h. The situation immediately prior to the collision is shown in Figure 3.1.

Why did the northbound train leave Rudstad before its scheduled departure time, and before the southbound train had arrived at Rudstad? Why was nobody able to notify the train drivers in time to avert the collision?

According to the traffic safety rules, the northbound train was to wait at Rudstad station until the scheduled time for departure and a green exit signal could be observed. The train driver did not know at what stations he would meet trains, and he was not obliged to ensure that the meeting train had arrived before he continued to the next station. The safety rules did not require the train guard to check the exit signals. Rather, the train guard was expected to concentrate on the safety of embarking and disembarking passengers. The train driver would activate a blinking light on the train to inform the train guard that he had a green exit signal.

The Åsta commission did not draw a firm conclusion as to whether or not a green exit signal could be observed as the northbound train departed from Rudstad. All technical traces were compatible with a steady red exit signal. However, due to the design of the interlock system, and because of inadequate logging of the safety system's operational status, the commission did not exclude the possibility of a technical malfunction of the signalling and safety system in connection with the accident. Based on technical evidence, interviews with passengers and analysis of the speed recorder, the commission concluded that the northbound train did not start inadvertently from Rudstad due to technical errors on the rolling stock.

The train movements on the Røros line could be followed on the screens at the Rail Traffic Control Centre in Hamar. Twenty seconds after the northbound train forced open a switch at the northern end of Rudstad station, the warning text 'S2-Switch control failed Rudstad' appeared a in 16 mm red font at the bottom of one of the screens. There was no acoustic alarm or blinking light associated with this warning. The rail traffic controller on duty was busy managing the traffic on another line, and did not monitor the displays of the Røros line continuously. He

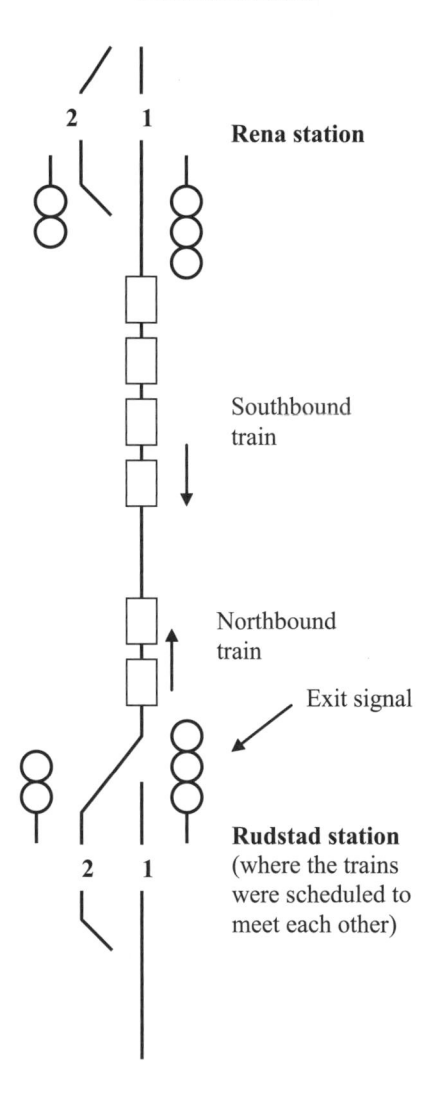

Figure 3.1 The situation immediately after the northbound train had left from track 1 at Rudstad, forcing open the switch at the northern exit of the station

discovered the warning text approximately three to four minutes after it appeared on the screen. He quickly understood the urgency of the situation and notified his colleagues at the traffic control centre immediately.

The Røros line employs diesel traction, so it was not possible to stop the trains by disrupting the electricity supply to the locomotives. The rail traffic controller on duty tried to contact the southbound train by mobile phone. He sought the mobile phone number of the southbound train, but was not able to find it on the telephone list for that day. Both trains had notified the traffic control centre of their phone numbers. However, the traffic controller on the previous shift had noted the numbers on the graphic timetable, but he did not record the numbers in the telephone list.

The rail traffic controller on duty then searched the list of locomotives, and found a phone number for the southbound train on that list. However, this number connected him to a wrong train. He tried the number of another locomotive which he knew was used on the Røros line, but he did not get an answer. One of the other rail traffic controllers obtained the correct mobile phone number to the southbound train by calling the operational centre and then the locomotive manager in Trondheim. However, when they called the southbound train at 13:17, about five minutes after the impact, they obtained no answer. One minute later, they received an emergency call from a mobile phone on the southbound train.

Could the Collision have been Prevented?

The account above describes how the defences that were in place on the Røros line failed to prevent the collision. However, we also need to consider whether other defences could or should have been implemented. The investigation commission identified several defences which might have prevented the collision if they had been in place (NOU 2000:30, 171–75):

1. There is a possibility that the driver of the northbound train would have waited at Rudstad if his schedule had contained information on where to expect meeting trains.
2. The presence of a train dispatcher at Rudstad would have reduced the risk that an error in the signalling system would lead to a critical situation. The driver of the northbound train would then have to obey signals given by the train dispatcher.
3. A different departure procedure, which required the conductor to independently check the exit signal, might have allowed a red signal to be observed, or prevented a transient green flash from being mistaken for a steady green signal.
4. An Automatic Train Control (ATC) system would probably have stopped the northbound train in front of the exit signal, or at least a short distance beyond the signal. The function of the ATC is to brake the train automatically in case the driver fails to observe a stop signal or a speed limitation. Beacons in the track transmit information on signal status and speed limitations to a computer in the locomotive. This computer will give an alarm and then activate automatic braking if the train driver fails to adapt the speed to an approaching stop signal or speed restriction.
5. An acoustic collision alarm in the train control centre at Hamar would have given the traffic controllers three to four minutes more time to notify the train drivers, provided that they reacted promptly to the alarm.

6. Rules as to how often the traffic controllers were to monitor their screens, combined with more staff at the control centre, might have reduced the time taken before an abnormal situation was noticed.
7. A train radio system would have allowed the traffic controller to reliably reach all trains on a railway section using a single number.
8. Procedures concerning recording of mobile phone numbers and hand-over between shifts at the Rail Traffic Control Centre might have prevented confusion as to the correct mobile phone numbers.
9. Procedures concerning information exchange between the operations centre in Trondheim and the Rail Traffic Control Centre in Hamar might also have enabled the traffic controllers to reach the southbound train in time to reduce the consequences of the accident.

The accident commission concluded that the Røros line lacked adequate barriers against single failure accidents.

The State of Traffic Control on the Røros Line

Was the Røros line particularly vulnerable with regard to train collisions? One way to answer this is to compare the Røros line at the time of the accident with the same line at a former point in time, for example 1950. Another way is to compare the Røros line to a Norwegian main line, where traffic is greater and the infrastructure better developed, for example the Dovre line. These comparisons are summarized in Table 3.1. The term 'crossing' here refers to the meeting of two trains running in opposite directions on a single track line. A 'crossing station' is a station on a single track line where two trains are scheduled to meet. If one train is delayed, a Rail Traffic Controller may decide to transfer the crossing to another station in order to prevent the delay from propagating to other trains.

Before 1990, train control on the Røros line was manual. Before a train could enter the single track section between two stations, the train dispatchers at the stations at both ends of the section reserved the section for that train by exchanging messages on the signal telegraph and recording the reservation in a book. The main conductor of the train then obtained a permission to drive from the train dispatcher at the departure station. The main conductor gave a departure signal to the train driver after he had received permission to drive and the train was ready to leave. The driver's timetable contained information about where he would meet trains, and the train drivers were required to check that the meeting trains had arrived before they started from these stations. This system contained few technical interlocks to prevent human errors. After a train had started from a station, there were no effective means to communicate with the train driver or stop the train before the train reached the entrance signal at the next station. However, a low rate of collisions was achieved by means of double-checking built into the

Table 3.1 Traffic control on the Røros line at the time of the Åsta accident compared to manual traffic control and traffic control on main lines

Safety Issue	Røros line 1950	Røros line January 2000	Dovre line 2000
Basic concept	Reservation and release of blocks performed by local train dispatchers communicating by signal telegraph. Transfer of crossings decided centrally by Rail Traffic Controller.	Centralised Train Control: Reservation and release of blocks performed by Rail Traffic Controller acting through decentralised interlock systems.	Centralised Train Control: Reservation and release of blocks performed by Rail Traffic Controller acting through decentralized interlock systems.
Information about crossing stations	Crossing stations marked in timetable. Driver notified if crossing is transferred to a different station.	Drivers not informed about crossing stations.	Drivers not informed about crossing stations.
Responsibility with regard to crossing trains	Train driver has independent obligation to make sure that the crossing train has arrived before he leaves station.	Train driver obeys signals; does not know the crossing stations.	Train driver obeys signals; does not know the crossing stations.
Departure procedure	Former departure procedure required train guard and driver to check exit signals independently. Train dispatcher at each station, present on platform when trains passed or departed.	New departure procedure: Driver only checks exit signal. Train dispatcher removed from most stations.	New departure procedure: Driver only checks exit signal. Train dispatcher removed from most stations.
Communication with train personnel	Signals.	Signals, block telephones and mobile phones.	Signals, block telephones, train radio and mobile phones.

Table 3.1 *Concluded*

Safety Issue	Røros line 1950	Røros line January 2000	Dovre line 2000
Emergency braking if driver fails to obey red signal	Two drivers in many trains formerly.	One driver only; Automatic Train Control not operative.	Automatic Train Control (ATC) brakes train automatically when approaching red signal.
Possibility of stopping trains from RTCC in emergencies	None with steam or diesel traction.	None (diesel traction).	Electric traction: Train controller may break electric power supply.

operating procedures. Moreover, many trains formerly had two drivers, so that a failure of one driver to observe a stop signal might be corrected by the other driver.

Starting in 1990, the Røros line was equipped with Centralized Traffic Control (CTC). This allowed signals and switches to be controlled and monitored from the Rail Traffic Control Centre in Hamar, removing the need for a train dispatcher at each station. Many opportunities for human error were removed, since the task of reserving and releasing rail sections, moving switches and setting signals was transferred to a technical interlock system based on fail-safe design principles. Similar remote control systems had already been installed on Norwegian main lines. However, due to the rather low traffic density on the Røros line, the technical solutions differed in some respects from the solutions used on main lines. The Røros line was not equipped with Automatic Train Control at the time of the Åsta accident, and trains would thus not be braked automatically if a driver failed to observe a stop signal. The Røros line was equipped with a less expensive safety interlock design than the main lines, and a different principle for detecting the movements of trains was employed. No train radio was installed, so mobile phone was the only available means to communicate with trains on their way between two stations.

At the time of the Åsta accident the Røros line had lost several of the barriers it had 50 years earlier, while it also lacked some of the barriers associated with fully equipped main lines. It seems highly plausible that the collision risk per crossing would be higher on the Røros line than at the Dovre line in January 2000, since the Dovre line had several barriers or recovery opportunities in addition to those on the Røros line. The number of barriers or recovery opportunities on the Røros line was higher in 1950 than in 2000. However, the barriers were different, since manual traffic control depends on extensive cross-checking among humans rather than on technical interlocks. We should thus not jump to the conclusion that the collision risk per crossing on the Røros line was lower in 1950 only based on the number of barriers and recovery opportunities.[1] However, the Røros line remained in a vulnerable transitional state in one important sense. A single erroneous action – a driver failing to observe a red exit signal – could lead to a collision between two trains. This state was also a deviation from a principle set down by the Norwegian State Railways in 1991 that all lines with remote control should be equipped with ATC.

The next section examines how this vulnerable state developed.

Evolution of Vulnerability on the Røros Line

The public commission traced several of the decision processes that contributed to the vulnerable state of the Røros line. The institutional and organizational context of these decision processes provides a background to the following outline of the decision processes related to procurement of ATC, change in the departure procedure, installation of an acoustic alarm at the Rail Traffic Control Centres, and the status of mobile phones.

The Institutional Context

The organization of railway operations in Norway was in a period of transition during the decade immediately prior to the Åsta accident. During most of the twentieth century, the Norwegian State Railway (NSB – Norges Statsbaner) was the dominating actor. Apart from tramways, commuter lines and a few minor branch lines, NSB had the roles of infrastructure owner, operator and regulatory body. NSB was thus not subject to external regulation of traffic safety. NSB was owned by the Norwegian state and was subordinated to the Ministry of Transport and Communications.

In 1996, the National Railway Administration (JBV – Jernbaneverket) and the Norwegian Railway Inspectorate (SJT – Statens Jernbanetilsyn) were established. The National Railway Administration owned the infrastructure and administrated its building and maintenance, operators' access to the tracks, as well as technical and operational regulations. The Norwegian Railway Inspectorate (SJT) was established in order to have an independent regulatory body. Both JBV and SJT

---

1 Human double-checking does not always lead to a great gain in reliability. For instance, two humans may be distracted by the same event. The double check of the exit signal apparently failed in the Tretten accident in 1975. In this accident, two trains collided under circumstances that resembled the Åsta accident, except that the departure procedure required the train guard as well as the driver to check the exit signal. On the other hand, it has been suggested that train drivers may have been more prone to fail to observe the exit signal before leaving a station in 2000 than in 1950 because they have adapted their driving behaviour to main lines with ATC.

were subordinated to the Ministry of Transport and Communications. After the reorganization, NSB was a national operator of passenger and cargo trains. NSB was still owned by the Norwegian state, but the organization was given a more autonomous status.

The Norwegian Railway Inspectorate had a small staff (10 persons in 2000, compared to 9,267 in the Norwegian State Railways). The Inspectorate had made clear to the Ministry that their personnel resources were not sufficient to fulfil the responsibilities of the Inspectorate. The Inspectorate concentrated its effort on system audits and emphasized that the responsibility for railway safety resided with the enterprises (infrastructure owners and operators). The Inspectorate based its work on the assumption that the conditions at NSB were approved and acceptable when the Inspectorate was established, and decided to concentrate on following up new systems and activities.

Procurement of Automatic Train Control

Installation of ATC (Automatic Train Control) on the Røros line had started at the time of the accident, and completion was foreseen in June 2000. However, ATC was not installed at Rudstad station at this time. The history of the ATC procurement process spans 25 years. The commission's report mentions several milestones in this decision process:

- 1975: A train erroneously leaves Tretten station and collides with a train which it is supposed to meet at Tretten. 37 persons are killed. The internal accident commission concludes that 'accidents like this one can probably only be avoided by implementing automatic forced breaking of trains driving past a stop signal'.
- 1980: ATC installation starts on Norwegian railways.
- 1990: In a report to the Ministry of Transport, Det norske Veritas, a major consultancy firm, recommends that ATC is installed on all lines with Centralized Traffic Control. A new accident at Lysaker station shows the importance of ATC.
- 1991: In response to a letter from the Ministry of Transport, the NSB administration states that it has been a precondition that all lines with Centralized Traffic Control should be equipped with ATC, and that this will be implemented by 1995.
- 1992: In response to a new request from the Ministry of Transport, NSB reconfirms that the Røros line will be equipped with ATC by the end of 1995. NSB and the Ministry propose that extra funds are allocated to ATC installation in the 1992 budget and again in the 1993 budget. These proposals were endorsed by Parliament.
- 1994: Installation of Centralized Traffic Control is completed on the Hamar–Røros section of the Røros line.

1995: The plan for ATC installation on the Røros line is submitted for comments. The traffic safety manager raises the issue of remote controlled sections without ATC at two top management meetings in NSB where the managing director was present.
1996: The plan for ATC installation on the Røros line is finalized. The traffic safety manager again raises the ATC issue in a memo.
1997: The National Railway Administration approves the plan for ATC installation on the Røros line. ATC installation on the Røros line was not given priority in the Norwegian Railway Plan 1998–2007.
2000: Åsta accident. ATC is partly operative on the Røros line.

This history spans 25 years. The need for ATC on sections with remote control was apparently accepted by the organization, and this understanding was reinforced through repeated reminders from the safety manager. However, the procurement process for ATC on the Røros line was very slow, and the window of opportunity created by the extra funds allocated by Parliament in 1992 and 1993 was not exploited.

Additional detail emerged in discussions with persons who took part in the internal accident investigation. The initial plan for installation of Centralized Traffic Control on the Røros line included installation of ATC. However, the funding for ATC installation on the Røros line was reallocated to other purposes. The traffic safety director at the time reluctantly accepted this, since it implied a delay for one year only. However, soon after, the traffic safety director left his job. His successor was placed at a position lower down the organizational hierarchy, and thus had more difficult access to top management. Moreover, based on a favourable fatal accident record, top management had decided that they could turn their attention to challenges other than safety, such as punctuality.

The departure procedure In September 1997, the National Railway Administration implemented a new departure procedure for passenger trains. According to the new procedure, the train guard was to concentrate on the safety of passengers entering and leaving the train. They were no longer required to check the exit signals. The new departure procedure applied to the whole railway system, including railways without Automatic Train Control. The Norwegian State Railways and the National Railway Administration claimed that the new procedure would improve overall safety by reducing the risk to passengers entering or leaving the train. By giving the train driver the sole responsibility for attending to signals in all situations, it would eliminate ambiguities about responsibility. Moreover, the new procedure provided for harmonization with the practices in Denmark and Sweden. The new departure procedure would also allow some trains to be operated without a train guard.

These claims were contested by the Norwegian Railway Inspectorate. The Inspectorate initially refused to accept the new procedure for railways without Automatic Train Control (May 1997), since this would create a new class of situations where a single human error could cause a collision. The Inspectorate requested that the National Railway Administration perform a safety analysis of the new procedure.

The National Railway Administration implemented the new departure procedure without acceptance from the Inspectorate. The Inspectorate informed the Ministry of Transportation about the situation. The Ministry instructed the National Railway Administration to postpone implementation of the new procedure until it was accepted by the Inspectorate, and to provide the Inspectorate with the required analyses. The National Railway Administration repeated its claim that the new departure procedure represented an overall safety improvement, and did not revert to the old procedure. The Inspectorate then indicated that it might accept the new procedure on a provisional basis, provided that the National Railway Administration committed itself to an ambitious plan for ATC installation. The National Railway Administration replied by giving a plan that was not particularly ambitious. However, the Inspectorate did not follow up the issue further.

Acoustic alarm at the Rail Traffic Control Centres

The commission noted that the Rail Traffic Controllers at Hamar might have had three minutes more to act if they had been notified about the imminent collision by means of an acoustic collision alarm. After a similar collision at Tretten in 1975, this issue was discussed in NSB, and acoustic collision alarms were installed at two of the six Norwegian Rail Traffic Control Centres (Narvik and Bergen) in 1976/77. NSB obtained a price offer for installation of an acoustic alarm at Hamar, but did not follow up the offer. Installation of alarms was again recommended in a Human–Machine Interaction study carried out by Forskningsparken in 1990 and in a report from SINTEF, a contract research institute, issued in 1994 after investigation of a train collision at Nordstrand. Before the SINTEF report was issued, NSB obtained a new price offer for installation of acoustic alarms. NSB corporate management decided that alarm procurement should be followed up by a NSB Safety Forum, in which high-level managers and safety specialists participated. However, the Safety Forum was discontinued in 1994, and responsibility for follow-up was transferred to the NSB Service Division. Memos from 1995 and 1996 mention that a pilot project is under consideration, but no further action has been documented between 1994 and the Åsta accident.

Status of mobile phones

Safety considerations may have contributed in a paradoxical manner to the absence of a clear procedure for recording mobile phone numbers at the Rail Traffic Control Centre in Hamar. Mobile phones were not considered a safe means of communication for railway operations. Significant parts of the rail network are outside mobile phone coverage, leading to inadequate availability of mobile phone communication. Mobile phones at Norwegian railways were part of the open, public network, and thus not adequately protected against abuse by the general public. Finally, ordinary mobile phones do not provide reliable information on the location of the caller (position control).

The use of the train radio system and phones mounted at signal posts are subject to strict regulations, in order to maintain their status as safe means of communication. Because mobile phones did not have a similar status, their use was not regulated in the same manner. This may explain the lack of rigorous procedures for communicating and recording mobile phone numbers.

Vulnerability and the accident

Figure 3.2 summarizes conditions and events that contributed to the Åsta accident. The accident sequence is depicted from left to right at the bottom of the Figure. Causal and contributing factors are grouped into layers, ranging from operational procedures and practice via infrastructures to managerial decision processes. The arrows indicate possible causal impact: that is, the event or state at the end of the arrow *might* have turned out otherwise if the event or state at the start of the arrow had not occurred. Some of the connections are rather tentative. For instance, we do not know with certainty that the accident would have been prevented if a different departure procedure had required the train guard to check the exit signal. A similar accident occurred at Tretten in 1975, when the departure procedure included such a requirement.

Figure 3.2 shows that the events at the sharp end are only the end of a complex story, where some defences were removed without new defences being put in place.

Reframing the Causal Analysis

The public investigation reports that are generated after major accidents tell stories about failure and blameworthiness. Actions and conditions are labelled and interpreted with a view to their contribution to the disaster. Such stories may be instrumental in producing change in the organizations involved. Confrontation with failure may help unfreeze dysfunctional basic assumptions (Schein 1997) and trigger a search for new practices.

However, the scope for learning may be expanded if we can show that the lessons from the disasters are relevant to other organizations. In order to do this, we may need to subsume the specific events and conditions in a given accident under more general concepts that are applicable to a broad range of organizations and activities. We may also need to reframe some of the apparently irrational and dysfunctional actions and practices described in the investigation report. Actions and practices that stand out as dysfunctional or irrational in the context of a disaster may be seen as rational or functional if the perspective is slightly

Derailed Decisions 67

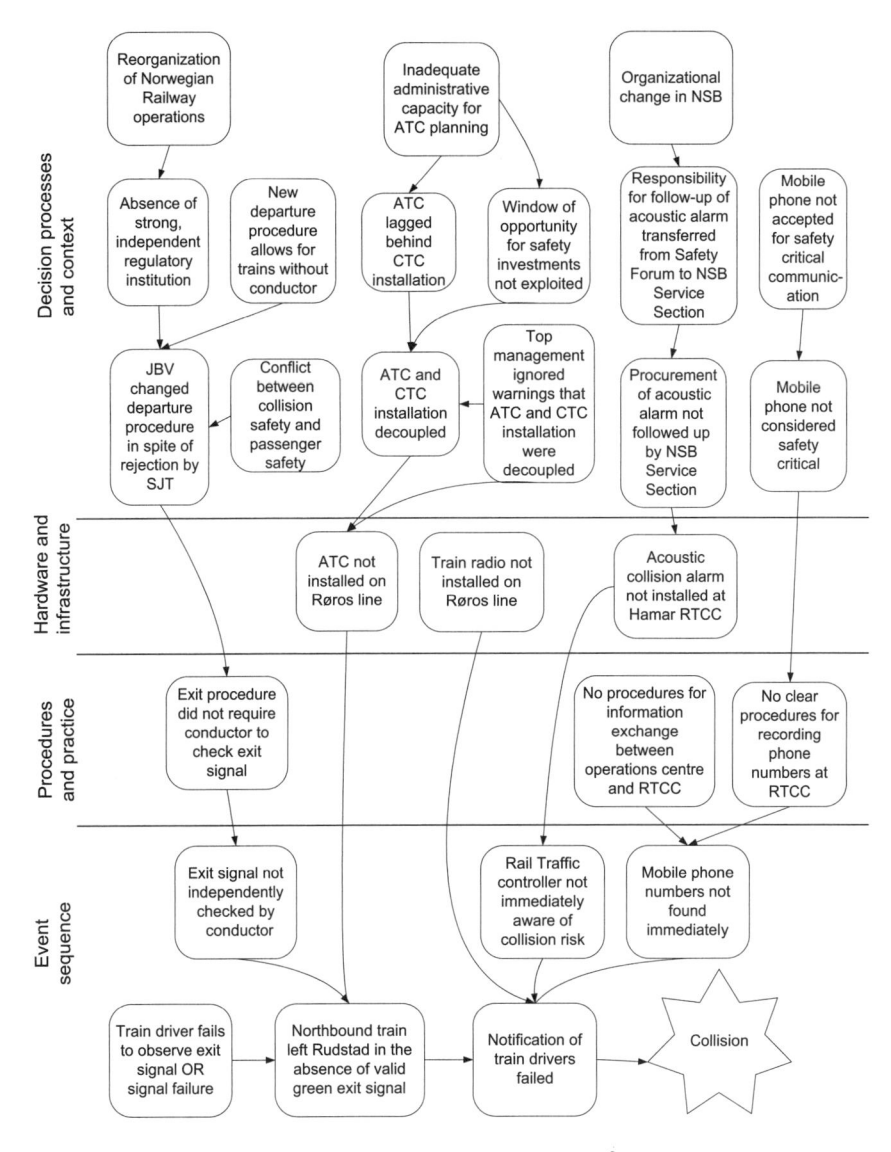

Figure 3.2   Causal and contributory factors in the Åsta accident

changed. A reframing may thus help us see how normal organizational processes can contribute to a disaster.

By extracting features of the decision processes that contributed to the Åsta accident, we can build a more general model of how some systems may evolve towards increased vulnerability. This draws on some of the literature on adaptation and drift, and a few pertinent works on organizational decision-making.

Migration and the boundaries of acceptable performance

As Wackers (Chapter 4) also points out, Rasmussen (1994; 1997) proposed that human activities at work are characterized by continuous adaptive search in the face of partially conflicting pressures and needs. Humans strive to keep the workload at a comfortable level, to find some intellectual joy in their work, and to avoid failure. They face requirements and pressures arising from, for example, productivity and quality. The workplace usually allows them considerable freedom to try out different ways of handling these partially conflicting needs and constraints. This can be depicted as a space of freedom delimited by (1) the boundary of financially acceptable behaviour, (2) the boundary to unacceptable workload, and (3) the boundary of functionally acceptable behaviour with regard to risk. In seeking a viable adaptation, humans will explore a large part of this space of freedom. Both the 'effort gradient' (related to workload) and the 'cost gradient' (related to productivity requirements) are likely to drive the activities towards the boundary of safe (that is, functionally acceptable) performance. An error or accident may occur if the crossing of this boundary is irreversible.

This model directs our attention to what happens at the boundary of safe performance. Is the boundary 'visible', that is, easy to recognize for the actors? (For discussion on the importance of making both work, boundaries and constraints visible, see Chapter 1.) What happens when an activity approaches or crosses the boundary? Will the actors receive an insistent warning from the system and have the opportunity to reverse their actions? Since many dangerous situations do not lead to disaster, is there a risk that they will adapt to warnings over time?

Norwegian railway operations take place in a context of conflicting demands from the environment. NSB was under pressure to reduce operating costs and operating deficits, and at the same time improve capacity and punctuality. Safety management in Norwegian railway operations has traditionally taken a reactive approach, with a focus on detailed operational rules (NOU 2000:30, 142). Managerial action to improve safety was typically taken in response to incidents and accidents, whereas it was uncommon to systematically assess whether a proposed change might jeopardize safety, or whether a given activity or system was safe. This approach may have led to a lack of clear and compelling criteria to identify the boundaries for acceptable risk in decision situations that are not covered by the operational rules.

The following excerpt from the investigation report (p. 153) seems to illustrate this lack of criteria to determine what is safe enough:

> The managing director of NSB at that time had claimed that it was a clear judgement in the organization that they had a safe and good system. On a question from the commission about whether he considered the safety on the Røros line on the 4th of January 2000 adequate, Ueland explained that the issue of safety was simple to him; it was either safe to drive trains, and then the trains

would roll, or otherwise the trains stood still. He claimed that he, like many others, had been living in the belief that it was safe to drive on the Røros line.

This lack of clear and compelling criteria may have made it difficult to make a case for changing decisions that could threaten safety.[2]

Distributed decision-making and decoupling

In its simplest form, Rasmussen's model of migration involves only one actor. A system is characterized by 'distributed decision-making' to the extent that it lacks a centralized decision-maker and each decision-maker has a model and information of a limited part of the problem (Brehmer 1991). In many complex systems, several activities take place in parallel. At a given moment, each actor may have incomplete or inaccurate knowledge about the state of the system and the ongoing activities. Moreover, the parallel activities may interact in non-obvious manners. The actions of one actor may displace the boundary of safe performance for another actor.

In this case, actors may strive for local optimization, based on their incomplete knowledge about the system. They will take into account the dangers and potential scenarios they know about, but not those that are 'invisible' from their local point of view. Actors may thus run risks, unless they keep an excessive margin around the perceived boundary of safe performance. Rasmussen and Svedung (2000) suggested that many accidents occur because unexpected interactions occur between activities which are usually not coupled in any functional way during daily work.

In routine operations on Norwegian railways, this accident potential is handled by enforcing detailed operational rules. A Rail Traffic Controller can maintain an accurate mental model of what happens along a line provided that train drivers, line workers and train dispatchers work according to the rules.

However, when it comes to managerial decision processes related to installation of Automatic Train Control and the new departure procedure, we can observe a decoupling of decisions that were strongly related in their impact on safety. The installation of Centralized Traffic Control (CTC) and Automatic Train Control (ATC) was not coordinated on the Røros line, in spite of a statement by the Norwegian State Railways in 1991 that all lines with CTC should be equipped with ATC. In a similar manner, the decision to change the departure procedure was decoupled from the issue of installation of ATC.

---

2   The National Railway Administration decided to revert to a departure procedure that involves double-checking of the exit signal by the train guard. This decision was made with reference to the (then) current railway safety regulations, which require railway operations to be planned, organized and executed with a view to avoiding that a single failure or erroneous action can lead to an accident with fatalities or serious injuries. This principle was established in 1999.

It is not clear why such decoupling occurred, but it did happen in spite of insistent warnings. The traffic safety manager repeatedly warned against the lack of ATC on the Røros line, and the Norwegian Railway Inspectorate insisted that changes in the departure procedure should not be decoupled from installation of ATC.

Breakdown in information flow and learning

Most major disasters are perceived as fundamental surprises by the organizations involved. However, several precursors or warnings are nearly always identified in hindsight by the media or accident investigators. This paradox is at the heart of Barry Turner's theory of 'man-made' disasters (Turner 1978; Turner and Pidgeon 1997; Pidgeon and O'Leary 2000). Turner proposed that accidents or disasters develop through a long chain of events, leading back to root causes like lack of information flow and misperception among individuals. During this 'incubation period', information and interpretations of hazard signals fail, and chains of discrepant events develop and accumulate unnoticed, often over years. Then, if someone acts in response to the signals, it often results in what Turner labelled 'the decoy phenomenon'. This is action taken to deal with a perceived problem which, in hindsight, is found to distract attention from the problems that actually caused the trouble. In many cases the company disregards complaints from outsiders and fails to disseminate and analyze pertinent information.

In the case of the Åsta accident, there was clearly somebody in the organization who knew about the dangers. The traffic safety manager had warned about the lack of Automatic Train Control on the Røros line at two different meetings in 1995 with the managing director of the Norwegian State Railways present. In 1996 the traffic safety manager issued a memo where he repeated his concerns. In 1997 he repeated his concerns in a new memo.

The Commission of inquiry asked the managing director of NSB at that time about his knowledge of the safety manager's concerns (NOU 2000:30, 153, author's translation):

> The managing director of NSB at that time ... could not remember that he had received [the memo from the traffic safety manager]. He explained to the Commission that he was confident that the consequences of [not giving priority to ATC installation] had been assessed, and said that there had been no disagreement in the organization about the reordering of priorities. He claimed that nobody in the organization had said that the priorities could not be changed, and that one could not postpone installation of ATC any longer. He further claimed that it was a clear judgement in the organization that they had a safe and good system. ... He claimed that he, like many others, had been living in the belief that it was safe to drive on the Røros line.

This excerpt illustrates the paradox which inspires the information processing perspective. The knowledge about the problem exists somewhere in the organization or its close environment, but this knowledge is not shared by the dominant decision-makers, and therefore not acted on.

How did this situation come into being? The traffic safety manager had recently been placed in a position lower down the organizational hierarchy, and thus had more difficult access to top management. At the same time, the confidence was developing in top management that the organization had tackled its safety challenges well enough to concentrate on other issues, such as punctuality and the development of new services. This conviction was founded on hard data – a favourable long time trend in fatal railway accidents, culminating in two very good years (1996 and 1997). There existed no strong external 'watchdog' who could effectively challenge this conviction.[3] This pattern fits well with Turner's notion of an incubation period where the organization systematically disregards warning signals. It also suggests that power relations play an important role in organizational information processing.

Interruption of problem-solving processes due to structural change

Structural change in organizations may lead to the disappearance of decision fora necessary to follow up specific problems. Like many organizations, the Norwegian State Railways went through extensive change processes during the 1990s. Structural change seems to have contributed to the failure to install acoustic alarms at some Rail Traffic Control Centres. After an accident at Nordstrand in 1993, installation of acoustic alarms was followed up by the Safety Forum (Sikkerhetsforum) at the Norwegian State Railways. The Safety Forum was discontinued in 1994 due to a restructuring of the organization. It was decided that the Traffic Section (Trafikksiden) of NSB should follow up the alarms. The Commission did not find any evidence that this task had been addressed. The decision process related to the acoustic alarm apparently lost its anchor in the organization when the Safety Forum was discontinued.

Decoupling of decision-making and action

We often think of decision-making and action as a single entity, assuming that decisions are a prerequisite to action, and that all decisions are geared towards action. This makes us think of a failure to carry out a decision as an anomaly. Brunsson (1989) proposed that decisions and actions should be analyzed as separate activities. Within this perspective, decisions can have different functions. A decision can represent a choice of a course of action which we believe to be the best alternative, but a

---

3   NSB was thus not subject to external regulation of traffic safety before the Norwegian Railway Inspectorate was established in 1996. The Inspectorate had very limited resources in its first years (NOU 2000:30).

decision can also be a means to discover or develop our preferences. Alternatively, a decision can be a means to mobilize, that is, to arouse commitment for coordinated collective action. Brunsson (1985) claimed that strong motivation is promoted by irrational decision processes, which give a positive picture of the action. A decision may also be a means to allocate responsibility, that is, to show clearly who the decision-makers are. Actors can acquire responsibility in the eyes of the world by dramatizing their decisions and making them look like choices. Finally, a decision may serve as an instrument for external legitimation and support.

Brunsson suggested that organizations can deal with inconsistent norms by acting according to certain norms and making their decisions according to others (1989: 188). Parliament may make a decision for the sake of ensuring the legitimacy of the institution and the public support to the political parties involved. Such a decision has to be clearly visible to the environment, but it need not be coupled to action.

It is not straightforward to determine in hindsight the function of a given decision within this scheme. For instance, the budget allocations for Automatic Train Control made by the Norwegian Parliament in 1992 and 1993 seem to make sense as a choice, but also as an attempt to allocate responsibility and as an instrument for legitimation of the decision-makers. We are thus not in a position to identify the functions or roles of specific decisions related to the Åsta accident. However, Brunsson's analysis points to the potential for decisions and actions to become decoupled when an organization seeks legitimation in the face of conflicting demands from the environment. Moreover, it also suggests that decision-makers will make their role less visible when they want to evade responsibility.

Incremental decision-making and the need for watchdogs

Lindblom (1959) suggested that public administrators often employ an incremental decision strategy when they have to tackle controversial and complex issues. Rather than facing a complex issue head on, through a comprehensive analysis where means and ends are kept apart, they build a policy by means of many small decisions based on limited analysis. This strategy of 'muddling through' enables the decision-maker to tackle problems that are so complex and controversial that they are difficult or impossible to handle by formal decision analysis techniques. However, because values are not explicitly addressed, important values may be lost unless there exist watchdogs with the requisite resources to promote those values. An organization that applies this strategy to decisions with a strong impact on safety may thus gradually drift towards greater risk if safety interests lack a sufficiently powerful watchdog. Thus, the absence of a strong, independent safety authority figures prominently in accounts of both the Challenger disaster and the Columbia disaster (Vaughan 1996, 2005; Presidential Commission 1986).

In the case of Norwegian railway operations, a powerful independent watchdog was missing. Until 1996, there was no independent regulatory institution. The Norwegian Railway Inspectorate was established in 1996, but the inspectorate had very limited personnel resources before the Åsta accident. The Norwegian Railway Inspectorate chose to assume that safety on Norwegian railways was acceptable when the inspectorate was established, in spite of the fact that the Norwegian State Railway had never been subjected to external control.

In 1997, the National Railway Administration changed the departure procedure on railway lines without Automatic Train Control in spite of an explicit rejection by the Norwegian Railway Inspectorate. The Inspectorate decided not to close down traffic on the relevant lines, which appeared to be the most relevant sanction available. The accident investigation commission suggested that the weak position of the Inspectorate was related to its being in the establishment phase.

Normalization of deviance through incident investigations

Drift towards higher risk may take place even when safety is handled in an explicit and systematic manner. The Challenger space shuttle was launched in unusually cold weather on 28 January 1986, in spite of serious concerns on the part of some of the design engineers that the o-ring seals between the joints of the solid rocket boosters might fail due to the low temperature. The concerns were borne out when Challenger exploded less than two minutes after launch, leading to the loss of seven astronauts and a serious setback for the space programme. Vaughan (1996) argued that the launch decision was the result of a long history of normalization of deviance. The behaviour of the complex booster joints was poorly understood. Design engineers and middle management were repeatedly confronted with unexpected leaks, burn-through and erosion of the o-ring seals. Each little surprise led the engineers to revise their rationale for risk acceptance. Through the accumulation of many small decisions, taken in a context of production pressure and economic scarcity, increasingly serious anomalies were defined as compatible with the rationale for risk acceptance. At the eve of the launch, the concerned design engineers found themselves caught by their previous rationales and unable to make a strong case for postponing the launch.

Could it be that similar processes took place with regard to incidents on Norwegian railways? A focus on operational rules may have contributed to normalization of deviance related to other aspects of safety. Incident investigations which emphasize deviations from operating rules tend to assign blame to the actors at the sharp end. Once blame has been assigned, the incident has apparently been accounted for, and there is no need to question other aspects of the system. The incident would not have happened if only the sharp end actor had followed the rules. Misdirected incident investigations may thus not only miss the point, but they may even reinforce trust in a deficient system.

A second look at the Åsta accident

We are now in a position to devise a second representation, using more generic concepts to capture conditions and mechanisms that may have contributed to the Åsta accident. The mechanisms at the sharp end are found at the bottom of Figure 3.3, whereas mechanisms associated with managerial processes are found higher up.

Starting at the sharp end, the Røros railway was vulnerable in the sense that a single erroneous action or technical failure could cause a disaster. The occasional failure of a train driver to observe a red signal is within the normal variability of human performance, even if the frequency may be less than one per ten thousand red signals encountered. In the absence of Automatic Train Control, collision alarms at the control centre and effective means of communication, such normal variability can develop into a disaster.

The decision processes that led to this vulnerability may seem irrational or irresponsible in retrospect. However, the apparent irrationality may also be viewed as an adaptation to contingencies that the decision-makers faced. Decision-makers facing conflicting demands often find it easier to build agreement and get things done if issues are faced one at a time. This may lead to the decoupling of decisions that interact in their impact on safety, as exemplified by the installation of Centralized Traffic Control and Automatic Train Control. The failure to follow-up on the procurement of acoustic alarms for the Rail Traffic Control Centres may be seen as poor management. However, similar failures may be quite common in organizations which undergo rapid structural change, since the fora or positions responsible for following-up may disappear during a reorganization. A failure to put all decisions into action may look like poor management. However, it is not uncommon that decisions and actions get decoupled when an organization faces conflicting demands from the environment.

Finally, the mechanisms that might have prevented drift towards a vulnerable state did not prove effective in the case of the Åsta accident. The Norwegian State Railways and the National Railway Administration relied on a reactive safety management strategy, and had not established clear criteria for risk acceptance in managerial decisions. Warnings from inside the organization, such as the memos warning against the lack of Automatic Train Control on the Røros line, were either ignored or suppressed on their way to top management. The external watchdog, the Norwegian Railway Inspectorate, was understaffed and lacked suitable means to enforce a decision.

Learning from the Åsta accident

What lessons can be drawn from the Åsta accident, if we take the perspective of other organizations handling complex and dangerous technologies in the face of conflicting demands from the environment? At the sharp end, the Åsta accident demonstrates the need for adequate defences in order to prevent the normal variability of humans and technology from causing disasters. At the blunt end, it demonstrates a need for mechanisms to prevent defences from eroding to a state where the risk becomes unacceptable.

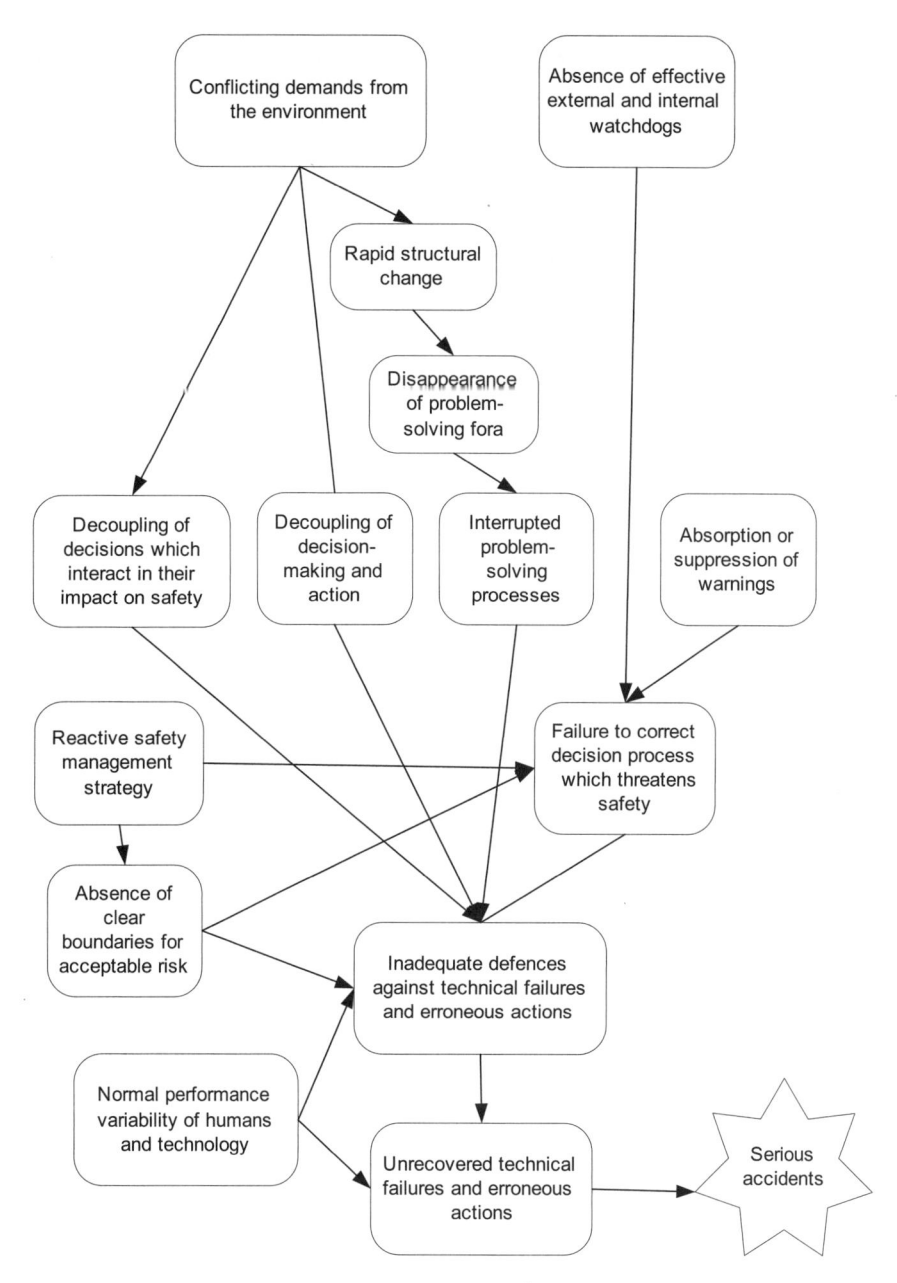

Figure 3.3 A second representation of the Åsta accident

Because the decision processes that contributed to the vulnerable state of the Røros line may seem irrational or irresponsible in retrospect, it is tempting to suggest that we need better decision-makers or better procedures for decision-making. However, if the apparent irrationality or irresponsibility is actually an adaptation to conflicting demands from the environment, then we have to realize that any complex organization facing conflicting demands may potentially drift into a vulnerable state. This implies that corrective mechanisms are needed if such drift is to be checked.

Efforts to improve safety in complex technological systems can be directed at several levels:

- Provide and monitor barriers and recovery opportunities.
- Educate operators and managers on the functioning and the vulnerabilities of barriers and recovery opportunities.
- Provide for adequate organizational learning from incidents.
- Monitor 'safety-critical' decision processes throughout the problem-solving cycle.
- Monitor organizational change with regard to effects on safety at the sharp end as well as blunt end problem-solving processes.
- Establish strong external counterpressures to counteract economic or other pressures that might cause the system to migrate towards a vulnerable state.

Information is not enough. A system will not necessarily stop migrating towards a vulnerable state just because somebody in the organization realizes that this may be happening. Counterpressures can come from several sources, such as members of the organization, clients, media or the general public. However, strong regulatory institutions may be necessary to maintain a steady and systematic counterpressure which is not limited to the most salient features of the last spectacular accident.

The human contribution to risk control

Incidents can often be interpreted as markers of both resilience and brittleness, depending on the perspective taken (see Chapter 2; Woods and Cook 2006). At first glance, the Åsta accident may look like yet another story about human failure, most prominently about management failure to take the necessary steps to prevent a single technical failure or erroneous action from leading to a serious accident. At the same time, our analysis of the accident indirectly sheds light on some of the mechanisms that contribute to the outstanding passenger safety records of railways.

The comparison between different modes of traffic control in Table 3.1 reveals an interesting role of humans on railway lines with manual traffic control. Humans, aided by operating rules, very simple communication tools, and a few technical interlocks, were able to operate railways with an impressive record of passenger safety. Only seven railway passengers were killed in 'train accidents' in Norway between 1854 and 1921[4] (Bergh 2004). Table 3.1 also suggests a partial explanation for the good safety record. The manual train control system contained numerous double-checks, which would capture many erroneous actions before they led to an accident.

However, rigorous safety rules are usually not enough to ensure safe operations. People have to follow the rules; at least they have to comply with the safety-critical rules most of the time. Several factors may have facilitated rule compliance on Norwegian railways (Bergh 2004; Gulowsen and Ryggvik 2004). Traffic safety had the status of a core activity and a core competence on Norwegian railways. The safety rules were not an add-on to the operating procedures; rather, the safety rules defined the operating procedures. Railway personnel were trained to understand the totality of traffic control, including the roles of other actors, and not only their own obligations. Moreover, contrary to what seems to be the case in some countries, results from a interview study indicate that Norwegian railway personnel find that the safety rules are realistic (with a few exceptions) and that the degree of detail is suitable (Guttormsen et al. 2003).

An incident on the Dovre line on 24 September 2000 illustrates the positive human contribution to safety on a modern main line with centralized traffic control. A northbound train had incorrectly been given a green exit signal at Hamar station and had entered a single track section already occupied by a southbound train approaching Hamar. The traffic controller at Hamar discovered that the two trains were on a collision course. Using the train radio, he was able to stop both trains in time to avoid a collision.

Skjerve et al. (2004) distinguished three ways in which humans at the sharp end contribute to a system's resilience against accidents. Defining 'safety barriers' as means of preventing a set of predefined unwanted events from occurring/and or reducing their consequences, they pointed out that humans may be included as barrier elements. In this role, humans will typically carry out a prescribed task using specific tools, such as when a train dispatcher uses the signal telegraph to notify the next station about a departing train.

Humans may also contribute to resilience by employing mindful work practices. These are discrete general safety-promoting work practices that may prevent the initiation of and/or interrupt unwanted but not explicitly predefined event sequences. This may be exemplified by defensive driving, where the driver acts on the assumption that something unexpected will happen. Instances of mindful work practices often go unnoticed, because their effect, if anything, is to cause nothing to happen. This may be the reason why mindful work practices are not readily found in the Åsta report. However, one instance of mindful work

---

4 The traffic volume was about 100 million passenger kilometers in 1890, and more than 800 million passenger kilometers in 1920 (Bergh 2004, 291). A major collision with six fatalities occurred in 1921. The term 'train accident' in the old statistics apparently did not cover instances where persons outside the train were killed.

practices at the managerial level deserves mentioning: The traffic safety manager warned executive management about the lack of ATC on the Røros line on several occasions.

The third way in which humans can contribute to resilience is by improvisation when faced with a dangerous situation that cannot be handled by following procedures. Such improvisation occurred when the rail traffic controllers at Hamar called the NSB operational centre to obtain the mobile phone number of the driver of the southbound train.

Humans serving as barrier elements may sometimes be replaced by hardware and software. This happened when Centralized Traffic Control was installed on the Røros line and automatic interlock systems took over many of the safety functions of the train dispatchers. In contrast, mindful work practices and improvisation represent uniquely human contributions to safety.

An afterthought on hindsight

Hindsight is an inevitable and pervasive aspect of all attempts to learn from experience. The implications of hindsight are paradoxical. Hindsight implies that we know more than the actors knew before the accident occurred. At the same time, we also know less, since it is very difficult to reconstruct the situation which the actors faced prior to the accident. The occurrence of an accident makes certain dangers and circumstances stand out in a way that may make us insensitive to all the other things that could have happened. We should also realize that some concepts may appear meaningful from a hindsight perspective but elusive in a foresight perspective. For instance, although boundaries of acceptable performance may be easy to define post hoc, with regard to a specific accident, it is not at all obvious how such boundaries can be identified based on foresight (Dekker 2006).

Knowing the facts also means that we can devise ad hoc explanations. We can, with a little ingenuity, make the facts fit in with our favourite conceptions of why accidents occur. This implies that accident analysis is not an effective way to subject our favourite theories to rigorous testing. These problems have a bearing on the explanations of the Åsta accident proposed in this chapter. I cannot prove that I have found the 'correct' explanations of the Åsta accident. However, I hope this analysis may contribute to our capacity to make sense of management action and inaction that has contributed to an accident.

References

Bergh, T. (2004), *Jernbanen i Norge 1854–2004* (Bergen: Vigmostad Bjørke).
Brehmer, B. (1991), 'Distributed Decision Making: Some Notes on the Literature', in Rasmussen, J. et al. (eds.) (1991), *Distributed Decision Making: Cognitive Models for Cooperative Work* (Chichester: John Wiley & Sons).
Brunsson, N. (1985), *The Irrational Organization* (Chichester: Wiley).

Brunsson, N. (1989), *The Organization of Hypocrisy: Talk, Decisions and Actions in Organisations* (Chichester: Wiley).
Dekker, S. (2006), 'Resilience Engineering: Chronicling the Emergence of Confused Consensus', in Hollnagel, E. et al (eds.) (2006), *Resilience Engineering: Concepts and Precepts* (Aldershot: Ashgate).
Gulowsen, J. and Ryggvik, H. (2004), *Jernbanen i Norge 1854–2004. Nye tider og gamle spor 1940-2004* (Bergen: Vigmostad & Bjørke).
Guttormsen, G. et al. (2003), *Utforming av Regelverk for Togframføring*. Report STF38 A03408 (Trondheim: SINTEF).
Hollnagel, E. (2004), *Barriers and Accident Prevention* (Aldershot: Ashgate).
Jernbaneverket (2000), *Togsammenstøt ved Åsta på Rørosbanen 04.01.2000. Systematisk Granskning ved Hjelp av STEP-metoden*. Rapport nr. 01/2000 (Oslo: Jernbaneverket, Uhellskommisjonen). [Report from the Accident Investigation Commission of the National Railway Administration.]
Lindblom, C.E. (1959), 'The Science of "Muddling Through"', *Public Administration Review* 19: 79–88.
NOU 2000:30: *Åsta- ulykken, 4. Januar 2000. Hovedrapport*. Justis- og politidepartementet. Statens forvaltningstjeneste, 2001. [The Åsta Accident, 4 January 2000. Main Report of the Public Commission of Inquiry.] <http://odin.dep.no/jd/norsk/publ/utredninger/NOU/012001-020007/index-dok000-b-n-a.html≥
Perrow, C. (1984), *Normal Accidents* (New York: Basic Books).
Pidgeon, N. and O'Leary, M. (2000), 'Man-made Disasters: Why Technology and Organizations (Sometimes) Fail', *Safety Science* 34: 15–30.
Presidential Commission on the Space Shuttle Challenger Accident (1986), *Report to the President by the Presidential Commission on the Space Shuttle Challenger Accident*. 5 vols. (Washington, DC: U.S. Government Printing Office).
Rasmussen, J. (1994), 'Risk Management, Adaptation, and Design for Safety', in Brehmer, B. and Sahlin, N. E. (eds.) (1994), *Future Risks and Risk Management* (Dordrecht: Kluwer Academic Publishers).
Rasmussen, J. (1997), 'Risk Management in a Dynamic Society: A Modelling Problem', *Safety Science* 27(2-3): 183–213.
Rasmussen, J. and Svedung, I. (2000), *Proactive Risk Management in a Dynamic Society* (Karlstad, Sweden: Swedish Rescue Services Agency).
Reason, J. (1997), *Managing the Risks of Organizational Accidents* (Aldershot: Ashgate).
Schein, E.H. (1997), *Organizational Culture and Leadership*, 2nd edition (San Francisco: Jossey-Bass).
Skjerve, A.B. et al. (2004), 'Human and Organizational Contributions to Safety Defences in Offshore Oil Production', in Spitzer, C. et al. (eds.) (2004), *Probabilistic Safety Assessment and Management*, vol 4 (London: Springer-Verlag).

Snook, S.A. (2000), *Friendly Fire: The Accidental Shootdown of U.S. Black Hawks over Northern Iraq* (Princeton, NJ: Princeton University Press).
Turner, B.A. (1978), *Man-made Disasters* (London: Wykeham Science Press).
Turner, B.A. and Pidgeon, N.F. (1997), *Man-made Disasters*, 2nd edition (London: Butterworth-Heinemann).
Vaughan, D. (1996), *The Challenger Launch Decision: Risky Technology, Culture, and Deviance and NASA* (Chicago: University of Chicago Press).
Vaughan, D. (2005), 'System Effects: On Slippery Slopes, Repeating Negative Patterns, and Learning from Mistakes', in Starbuck, W. and Farjoun, M. (eds.) (2005), *Organization at the Limit: NASA and the Columbia Disaster* (Malden, MA: Blackwell).
Weick, K.E. (1990), 'The Vulnerable System: An Analysis of the Tenerife Air Disaster', *Journal of Management* 16(3): 571–93.
Woods, D.D. and Cook, R.I. (2006), 'Incidents – Markers of Resilience of Brittleness?', in Hollnagel, E. et al. (eds.) (2006), *Resilience Engineering: Concepts and Precepts* (Aldershot: Ashgate).

Chapter 4

Offshore Vulnerability: The Limits of Design and the Ubiquity of the Recursive Process

Ger Wackers

Prologue: Loss of Functional System Integrity

[B]1991, Sleipner A:

On 23 August 1991, seismological stations in southern Norway registered a vibration with a force of 3 on the Richter scale. Investigations identified the cause of the vibration: a concrete offshore production platform (Sleipner A) that was under construction and being prepared for 'deck mating' sprang a leak, capsized, imploded on its way down and hit the seafloor as a mass of rubble and bent reinforcement steel in Gandsfjorden near Stavanger. No lives were lost. Sleipner A was number 12 in the series of 'gravity base structures' (GBS) of the Condeep-type build by Norwegian Contractors (NC). And it was a relatively small one. During the 1980s NC had built the Gullfaks-series, with the 262 meter high Gullfaks C being the largest Condeep ever built, in terms of volume of concrete. Construction of the tallest one, the 369 meter high Troll, had just started in July 1991. Sleipner A was designed to stand in 89 meters of water and would be 110 meters high.

[B]1997, Super Puma helicopter, LN-OPG:

On 8 September 1997, a AS 332L1 Super Puma helicopter, equipped with a vibration-based Health and Usage Monitoring System (HUMS) and operated by Helikopter Service AS (HS), crashed into the sea on its way to a floating production vessel at the Norne field, 200 km west of the Norwegian coast. None of the 12 people on board survived the sudden accident. The technical investigation conducted after the recovery of the engines identified a mechanical fatigue failure in a 'splined sleeve': a part connecting the shaft of an engine to the shaft of the gearbox. The HUMS sensor monitoring this connection was found to be 'unserviceable'; and it had been so for some time. Through manual retrieval and examination of HUMS data batches from previous flights HUMS engineers discovered a trend in one of the parameters. Although in no way certain, this opened up the possibility that if the defect sensor had been serviceable, it could have picked up the trend, which might have surfaced as an automatically generated alert and the accident could have been prevented.

[B]2004, Snorre A:

During the execution of a slot recovery program, on 28 November 2004, Statoil lost control over well P-31 A on Snorre A, a tension leg platform (TLP) on the Norwegian continental shelf. A scabliner was being pulled from a suspended, plugged well that had been opened to the reservoir again to facilitate injection of concrete into the reservoir section of the tubing. Pulling the scabliner exposed old holes in the tubing below seafloor level and sucked gas from the reservoir (swabbing). This resulted in a parallel gas blow-out with high-pressured gas escaping from craters in the seafloor and from the topside blow-out preventor (BOP). The worst case scenario entailed the possibility of loss of the platform as a result of either the loss of buoyancy, or due to loss of one or more of the anchors on the seafloor being undermined by the craters through which the gas erupted. The capsizing and sinking of the platform on site would have ruptured several if not all of 30 oil-carrying risers and would have created a massive oil blow-out from multiple oil wells on the seafloor in water 350 meters deep, covered by the steel wreck of the platform.

Introduction

In spite of all the efforts to design high-hazard–low-risk systems, large-scale accidents happen that signify a failure to maintain functional system integrity. In retrospect these system failures can be fruitfully analyzed in terms of the interaction of an unintended event with specific system vulnerabilities that were retained within the system or that developed over time without properly being recognized and acted upon. These vulnerabilities can arise in system design, in maintenance regimes, in operational practices or in the planning and execution of modification programmes. I have defined vulnerability (Wackers 2004) as 'a reduced ability to anticipate, resist, cope with or recover from "events" that threaten the achievement or maintenance of performative closure'. *Performative closure* is shorthand for the achievement and/or maintenance of core task completion while maintaining functional system integrity.

Over the past couple of years I have conducted a number of in-depth case studies of large-scale accidents in the North and Norwegian Sea offshore industry, involving very different kinds of systems, at different stages in their life cycle and involving design, maintenance and modification work. To counter or avoid a 'reduction to the particular' – the argument that because of their diversity the cases cannot be compared and no general lessons or insight can be derived from them – I will, in this chapter, elaborate a framework that can hold them all. Drawing on recent work in complexity theory and on insights from science and technology studies, I start with a couple of observations that have to do with the fact that fossil fuels by far are the most important source of energy on which our societies depend.

Self-sustaining Societies: Energy Hunger and 'Lifelines'

First, let us start with the observation that our complex and advanced societies exist and live far from thermo-dynamic equilibrium. Maintenance of the current level of organization and order requires the continuous throughput of large amounts of energy. They are what Prigogine called dissipative systems: systems that exhibit self-organization, forms of emergent order, as a result of the throughput of matter, energy and information (Nicolis and Prigogine 1989). Beyond a certain level of complexity they become self-sustaining. Our bodies have learned to tap the food we eat for the energy that is stored in it and to exploit that energy to maintain, far from (thermodynamic) equilibrium, the organizational integrity of the body. This in turn allows us to act on our own behalf in the world as (relatively) autonomous agents (Kauffman 2000, 4). Our lives – as self-sustaining entities – rely on that ability to utilize the energy stored in food. Like our bodies, requiring a regular supply of energy in the form of food, relative shortage of energy supply for our cities and societies generates a hunger for energy that has to be satisfied. Wars are being waged to secure the long-term supply of energy. It is possible to reduce the consumption of energy, to go on a diet, but it is not possible for societies to choose to retreat to a lower level of complexity and its associated lower requirement for energy. The loss of energy supply eventually leads to loss of order, disintegration and death.

Historically, the transitions that marked the successive steps from one level of complexity to another – the urban explosion, the industrial revolution – are inseparably linked with the development and deployment of technologies (like the plough, steam engine, rotary drill and internal combustion engine) that allowed for a substantial increase in the amount of energy to be extracted, transformed and used from natural sources (of storage) (Johnson 2001; Mumford 1961). Today, with more than 50 per cent of the world's population living in cities, humankind is a species of energy-consuming, automotive city dwellers.

Although the modern history of oil begins in the latter half of the nineteenth century, Yergin (1991, 13) argues that it is the twentieth century that was completely transformed by the advent of petroleum: the history of oil being intimately linked with the rise of capitalism and modern business and being intimately intertwined with national strategies and global politics and power. Energy has become a strategic resource, a factor in the rise and fall of economies and civilizations. Human history, Roberts (2004) argues, may well have been marked by a series of energy crises that either killed off a particular civilization or helped push it to the next level of technological and economic development; to the next level of complexity.

The world experienced the disruptive nature of a shortage of energy when in 1973 OPEC curtailed the production of oil by its members. This boosted the development of the North Sea as an oil province, from which both Great Britain and Norway profited. According to Upton (1996, 1):

the North Sea oil and gas industry came to Britain in the 1970s and 1980s as manna from heaven. It fed the nation's economy and sustained its hopes during one of the darkest periods of recent British history. The last traces of the British empire had finally gone. An energy crisis seemed set to last for ever. Wrenching financial upheavals finally reduced Britain from world leader to pauper, forced to borrow from the International Monetary Fund on the humiliating terms more normally reserved for third-world debtors.

The offshore installations out there in the North and Norwegian Sea are our *lifelines*. No more, no less. They are the umbilical cords that sustain our society's energy-intensive, urbanized way of being.

Framing Energy Supply as Tradable Commodities, Financial Interests and Risks

The previous section suggests a second observation. The production of energy in quantities sufficient to sustain a society at its current level of complexity is intimately linked with, or better, *framed* by and *entrained* in the global flow of capital, the trading of financial interests and the management of financial risks. According to Callon (1998) the process of *framing* constitutes the field; it shapes interactions and contracts between actors and companies; it shapes experiences, provides cognitive repertoires and suggests courses of action. Borrowing the concept of framing from Goffman (1974), Callon (1998, 249) argues that:

> [t]he frame establishes a boundary within which interactions – the significance and content of which are self-evident to the protagonists – take place more or less independently of their surrounding context. ... It presupposes actors who are bringing to bear cognitive resources as well as forms of behaviour and strategies which have been shaped and structured by previous experience.

'The framing process', Callon (1998, 249) goes on, 'does not just depend on this commitment by the actors themselves; it is rooted in various physical and organizational devices' and a 'metrological infrastructure' (statistical tools, formats for data extraction and collection, establishment of databases, etc.).

The notion of *entrainment* suggests that one process – in our case the safety-critical work of designing, operating, maintaining and modifying the complex technological systems required for the production of energy – is entrapped in or caught by the dynamic of another process that is guided by the premises, values and principles of economic thinking, financial markets and optimal business performance (Ancona and Chong 1996).

By the seventeenth century, traders and business operators had invented ways of generating amounts of investment capital too large for a single person or company to bear. Investors were allowed to own a share or stock of the company, earning the right to take part in the profit, but also sharing in possible losses. Furthermore, these financial interests were tradable. Mutual organizations – stock exchanges – were set up to facilitate, standardize and regulate this trading of financial interests. To finance expensive and risky trading expeditions to the Far East, in 1602 the Dutch East India Company issued the first stocks on the Amsterdam Stock Exchange, establishing itself as a joint stock company. Hence, these solutions were already in place to finance the capital-intensive and risky development of coal mines in the eighteenth century and of oil and gas fields in the nineteenth century. Today, all major oil (or better: energy) companies are listed on the world's stock exchanges where small pieces of ownership in the companies are traded every day.

The production and supply of energy are framed by and entrained in financial markets because coal, oil and gas are tradable commodities. Agreements or contracts to sell and buy oil or energy at a future time, at a pre-set price, are tradable too, through the London-based International Petroleum Exchange, founded in 1980. Details of the workings of stock and futures and options exchanges do not have to concern us here, though it is important to note that it requires specialized training and knowledge, skills and a dedicated infrastructure of tools, databases and accounting systems. And it is a business that is efficient, flexible and fast. This fast temporal logic of the operation of financial markets is important, because it is at odds with the much slower temporal logic of safety-critical processes required for reliable design, operation, maintenance and modification of complex technological systems (see next section). The influence of the fast dynamic of financial markets does not stop at the boundaries of the company. Upper-level management is very much concerned with the maximization of shareholder value through the improvement of cash-flow generating capacity and cost reductions. These are adaptive processes that optimize business performance. Expressions for the quality of the management's work derive from this field. The fast dynamic is translated down to lower-level company units and to subcontractors through incentives, formalized in contracts, that favour meeting deadlines, productivity goals and operational effectiveness (uptime). These are the time-pricing mechanisms that turn time into money. The particular, but contingent, framing of offshore production installations and helicopters as assets, as cash flow and profit generators and as various forms of costs in the company's accountancy system – self-evident to the protagonists – is accompanied by a cognitive *imaginative deficit*: an inability to imagine the vulnerability inducing consequences of the entrainment of safety-critical work in a corporate culture of business performance optimization (Adamski and Westrum 2003). That is why large-scale accidents (be)come unexpected, unimagined: the impossible happening.

The Temporal Logic of Reliable Energy Production

Now we are ready to make a third observation. There are adaptive processes that are large and slow – with brief accelerations in transition periods – and there are adaptive processes that are small and fast. The daily trading of company stocks and futures, responsive as it is to all kinds of events that occur in the world (hurricanes in the Caribbean and cold winters, the outcome of elections and military operations), is an example of the latter. The outcomes of this process can be traced on a daily and even hourly basis in the changing value of company shares on the stock exchange and in the volatile price of oil on the International Petroleum Exchange.

The processes underlying the physical production of oil and gas live on longer timescales. The formation of oil and gas reservoirs must be measured on geological timescales. The time it took to negotiate the international agreements establishing the offshore boundaries between the territories of the various states in the North Sea (signed in Geneva in 1958) and the states' ownership of natural resources discovered in the seafloor within these boundaries, can be measured in decades. In the high north this process is not yet completed, the boundary between the Norwegian and Russian continental shelf in the Barents Sea still being a matter of dispute. It takes years to design, engineer, build and commission the installations that have to produce the oil and gas from the geological reservoirs. They are typically designed to be operational for decades, depending on production prognoses derived from the size of the reservoir and available production technology. However, during their lifetime they will be modified several times as new technology or new technologies are introduced. Underlying the production of oil and gas are the state's laws regulating the mining of fossil fuels by private or state companies, regulations issued by the state's inspectorate, and industry standards embodying the collective experience and best practices of a whole branch of industry. These are processes that live on timescales of different lengths. It is not that they do not change, but that they change at a slower rate. In complex systems these adaptive processes living on different timescales co-exist, they co-evolve and influence each other mutually. They are nested into each other, constituting a dynamic structure, rather than a rigid hierarchy. And there is a temporal logic inherent in them. Fast changes in slow processes can be very disruptive.

This framework derives from complex theoretical work being done to better understand adaptive, evolutionary changes in ecosystems as well as in economies and social systems (Gunderson and Holling 2002). Gunderson, Holling and others use the term 'panarchy' (as being distinct from hierarchy) to refer to these complex, self-organizing, multi-level but nested systems. Such a term provides tools for thinking about biological organisms and ecosystems, as well as for thinking about human organizations at different levels of aggregation and the temporal relationships between them.

The fast levels invent, experiment and test; the slower levels stabilize and conserve accumulated memory of past successful, surviving experiments. The whole system is both creative and conserving. The interactions between process levels combines learning with continuity, change with stability. Slower levels set the conditions within which faster and slower ones function, embodying the 'memory' of past experiences. The faster levels explore and test the constraining boundaries in their local, adaptive and creative search for adjacent possibilities (for alternative discussions about how organizations navigate boundaries see also Chapters 2 and 3).

Gunderson and Holling develop an interactive loop between the mechanisms of 'revolt' and 'remember'. 'Revolt' is the label that Gunderson and Holling (2002) attach to the influence that the faster levels exert on the slower and larger levels. The mutual interactions between process levels that Gunderson and Holling call 'remember' and 'revolt' are mechanisms through which the system coordinates the behaviour of its various parts and achieves a measure of temporal alignment or synchronization. It is interesting to note that Marion (1999, 18, 74) uses the terms 'resonance' or 'correlation' for the interactive, coordinative mechanism through which the behaviours of different actors achieve a measure of sync-ness. All process levels have mechanisms to evaluate the adaptive changes of the smaller and faster level constituting a selection environment for these changes, retaining those that were successful, resulting in improved fitness for the system as a whole.

This framework can serve as a heuristic tool to think about the design of sustainable complex systems, that is about how to design the feed-forward mechanisms (that embody collective experience) and the feedback mechanisms (that embody creative solutions to local problems). However, the framework can also serve as a heuristic to think about how systems can become vulnerable and eventually fail because they are no longer able to achieve or maintain 'performative closure' (Wackers 2004), as they are no longer able to coordinate key processes in situations that require a high degree of integration. For the purpose of thinking about vulnerability we can extrapolate from the ideas of Gunderson and Holling, thinking about their feed-forward loop label 'remember' in terms of 'design' and the feed-back loop label 'revolt' as the 'regularity gradient'.

The notion of regularity gradient was introduced in the analysis of the 1997 Norne helicopter accident, in an attempt to grasp exactly the influence exerted by the fast-moving dynamic of the oil and stock markets on the reliability and vulnerability of safety-critical work performed at work floor level that was shaped, but not completely determined, by regulations and standards that had their origin in the slower processes of engineering departments (Wackers and Kørte 2003; Wackers 2002).

It is important to note that the (Norwegian) offshore industry is not 'designed' in accordance with such complex system principles. The industry employs a rational and instrumental engineering notion of design.

Rational Design, Long-distance Control and the Distribution of Responsibility

Slow processes set the conditions for, but do not determine, the fast processes. The state owns the natural resources (including oil and gas) in the underground of its territory. As the legitimate authority the state can grant licences to oil companies. In offshore oil and gas fields the licence to produce is always granted to a group of companies (the state itself, represented by another of its 'organs', can be among those legal entities to whom the licence is granted). The granting of a field licence can be conceived of as an economic transaction that is constitutive of a discretionary space for an actor (see Coeckelbergh and Wackers 2007 for a discussion).

The 'power of free decision' is not absolutely free, however. The granting of discretionary power is always conditional; the discretionary space is bounded. The distribution of discretionary power by a legitimate authority establishes the right of the actor: (a) to ownership of some property; (b) to exercise (or not use) that discretionary power and (c) to freedom from undue interference (that is, from the public, environmental groups or the media – see Lakoff and Johnson 1999 for further discussion). The latter also comprises freedom from undue interference from the state itself. Hence, the granting of a licence itself constrains the state's possibilities to interfere with the way in which oil companies operate, as long as this is done in accordance with the stated purpose, licence conditions and without causing detrimental effects for human and non-human others.

The state now faces a problem of long-distance control: how to ensure that the high-risk technologies out there, in the form of exploration and production installations, perform reliably. The state tackles this problem through the development of a set of functional rules and regulations. The regulations are functional in the sense that they specify the goals (or performance criteria) that have to be achieved, but they do not specify how – the means through which – these goals have to be achieved. For example, the regulations specify that there shall always be two well barriers available during all well activities and operations where a pressure differential exists that may cause uncontrolled outflow from the borehole/well to the external environment. Industry standards like the NORSOK standard (2004) elaborate how this can be achieved technically, during various kinds of well operations and transitions between them, embodying the industry's collective experience and standards of best practice.

The set of functional rules and regulations is stacked, consisting of several layers that are 'owned' by different institutions. Government and parliament own the slowest level of the Petroleum Act. The Petroleum Safety Authority (PSA), an organ of the state established to monitor the oil companies' compliance with the rules and regulations, develops and owns a set of general and more specific prescriptive regulations, including criteria that must be met by offshore constructions in concrete, or procedures for the planning and execution of modification programmes. Industry standards such as the NORSOK standards constitute the next level. Finally, oil companies have internal requirements and guiding documents that attempt to control and regulate the work of organizational sub-units such as project planning teams or platform-specific operations units.

A number of infrastructures that gather 'outcome' data – that can be fed into quantitative reliability and risk equations – and feedback mechanisms are in place to learn from offshore experience. Maintenance data, for example about equipment failure rates, are linked to design and modification planning. Reports on deviations, incidents and accidents are analyzed and may lead to recommendations for changes or clarifications in the set of rules and regulations. Incident investigations can be conducted by the company or by the Petroleum Safety Authority. The regulations comprise rules about reporting procedures.

People working within this rational design frame – in Callon's (1998) sense – hold the conviction that in-principle compliance with the set of rules and regulations makes reliable operation of offshore installations possible. The rules are not perfect yet, still evolving and learning, but the approach as such is not questioned. Law (1994) speaks about different modes of ordering. This ordering mode is exemplified by rational technical system design and defence-in-depth design philosophies (see Reason 1997 for further discussion). It gives rise to a vocabulary of operational and organizational barriers being added to the two-tested-barrier requirements of geophysical engineering.[1] It also gives rise to the implementation, monitoring and enforcement of a multi-level set of functional regulations and industry standards that provide practical grammars for a range of specific situations. The mode of ordering through design also gives rise to the design and implementation of reliability and safety information systems that gather and process outcome-related data. This is the dominant way of thinking that frames the view on industrial installations and their operation in the offshore industry: a way of framing that is firmly grounded in regulations, in methods of quantitative risk assessment and in information systems. But it is also grounded in the slow process level of the Petroleum Act.

Norway's Petroleum Act attributes legal responsibility for offshore operations exclusively to the oil company that the state appoints as operator on a field. As a result, the state restricts the distribution of responsibility – that should be shared among all companies that own shares in a field licence – to just one company that assumes operational responsibility and liability. Other partners in the field licence are exempted from legal liability. Here issues of technical integrity and operational reliability of offshore installations are partially divorced from the (financial) management of a company's financial interests. Concerns about technical reliability and vulnerability are deferred and delegated to the operating company, where they are delegated further to technical organizational sub-units. The Petroleum Act also sets and constrains the mandate of the PSA, restricting its formal investigative powers to applicable rules and regulations and to the conditions specified in the specific licence to operate.[2] The accident investigation methodology employed by the PSA – called Man, Technology and Organization – follows these constraints set by the Petroleum Act and its mandate. Being sensitive to the interplay and frictions between man, technology and organization, the methodology restricts its organizational scope to the immediate operational procedures surrounding the technological system that failed. Business deals between oil companies, financial

---

1   For a review of the concept of barrier in relation to vulnerability see Rosness et al. 2002.

2   Some commentators claim that the PSA should relate to these regulations and licence condition only, and not develop an opinion about anything that goes beyond them.

incentive schemes for line managers or economic incentives in contracts between the operating company and contractors fall outside its scope.

Regularity Gradient

Reason (1997, 16) advises his readers to adopt an accident causation model that is confined to the manageable boundaries of the organization concerned. However, organizations are open systems and events and processes in an organization's environment do not stop at its boundaries but may be translated in various ways to company internal processes. To stop at the company's boundary in the investigation of causal chains is therefore arbitrary. Tracing the sources of system vulnerabilities – leading to large-scale accidents – beyond organizational boundaries inevitably brought me to the fast-moving dynamics of the oil and stock markets. And to the ways in which oil companies accommodated 'market stress' and translated it down the system (Wackers and Kørte 2003; Wackers 2002; 2004; 2006).

This market stress had the face of a dramatic drop in oil prices. After a period with high oil prices, in the range of $60–90 per barrel, from the early 1970s (following OPEC's production reductions), oil prices dropped dramatically to levels in the range of $30–40 per barrel in the mid 1980s: the years in which Statoil and Norwegian contractors were negotiating the contract for a new Condeep-platform, Sleipner A. Following a period of about a decade with unstable price levels in this range, the price of oil dropped further in the second half of the 1990s to levels in the range of $10 per barrel: the years in which Statoil and other oil companies renegotiated their contracts with helicopter operating companies such as Helikopter Service, the Norne Helicopter accident occurring in 1997. In 1999, the year in which Norsk Hydro and Statoil bought Saga Petroleum – and Hydro took over as operating company on Snorre A – the oil price slid under the level of $10 per barrel. Dropping oil prices induce in oil companies acute cash-flow problems. For the financial health of a company, its ability to generate and maintain a cash flow that is sufficient to pay bills and salaries is a more important indicator than the profit it makes. With cash-flow generation being compromised, one way to adapt to the reduction of income from the sale of oil and gas is to reduce cost levels.

Sleipner A was going to be the first production installation on the Sleipner gas field. The history of the Sleipner A project demonstrates various ways in which Statoil tried to reduce or limit the total costs of the installation. This resonated positively with the drive of Norwegian Contractor (NC) engineers to optimize the Condeep design. Local optimizing design adaptations performed by the design team were evaluated positively by the slower level of the NC's organizational environment – its senior engineers conducting internal engineering reviews – but they also passed the test of external quality control procedures. Sleipner A failed the final test in the physical selection environment of Gandsfjorden near Stavanger, during a rehearsal of movements in preparation for deck mating (Wackers 2004).

In the 1990s oil companies renegotiated their contracts with helicopter operating companies. They introduced vibration-based in-flight monitoring systems called HUMS as a requirement into contract negotiations at a stage in which the prototypes that had been tested in British field trials had not proven their early diagnostic capacity. The HUMS system monitors changes in vibration patterns during flight, storing data batches that after return to the base are downloaded and analyzed for trends by specific software. In its design it is an early diagnostic tool that tries to detect imminent mechanical failure as early as possible so that corrective maintenance can solve the problem before an accident occurs.

The new contract negotiations resulted in a reduction of the helicopter hour price of 50 per cent. Of course, this induced cash-flow problems in Helikopter Service that were accommodated by increasing the number of flight hours per helicopter per day. Maintenance work was displaced to evening and night hours and to weekends. Helicopter capacity and pilot capacity was already stretched to a maximum when Statoil demanded extra flights to the Norne ship in order to complete topside work. The completion of topside offshore – instead of in the dock or near land – and simultaneously to perform the work to hook the ship up with the production wells, was an adaptive move in order to meet production start-up at a specific date to which Statoil had committed itself through already negotiated and signed sales contracts (futures). With false positives in the range of 198 out of 200, the HUMS output was still very unreliable. An unserviceable sensor of a non-mandatory and unreliable monitoring system would certainly be insufficient to ground a helicopter. If the sensor had been serviceable, at least there would have been a possibility for it to have captured the imminent mechanical failure that caused the helicopter to crash in 1997, en route to the Norne ship (Wackers and Kørte 2003).

Taking a lead from Rasmussen's work on gradients in work spaces (1994), the notion of regularity gradient captures these influences from oil market dynamics on the performative capacity of technological systems. Rasmussen (1994) limited the operation of his notions of an 'effort gradient' and a 'cost gradient' to the work space of humans working at the interface with technical systems. Regularity gradient expands the notion of a gradient to encompass interactions between workspaces and organizations. The advantage of the term gradient is that it carries no deterministic overtones. There is no push or pull. There is only a difference that might be described in terms of energy levels or levels of resistance, suggesting pathways for movement or directions of flow.

The notion of regularity has to do with an organization's ability to meet demands for delivery (Østebø and Grødem 1998, 1101). In the offshore industry; that is, in the business of meeting society's growing energy demands, this ability is reflected in the value of company stocks on the stock exchange, where investors expect a return on investments.

Reason (1997, 16) argued that:

> [t]he economic and societal shortcomings ... are beyond the reach of system managers. From their perspective such problems are given and immutable. But

our main interest must be in the changeable and the controllable. For these reasons, and because the quantity and the reliability of the relevant information will deteriorate rapidly with increasing distance from the event itself, the accident causation model presented ... must, of necessity, be confined largely to the manageable boundaries of the organization concerned.

We argued to the contrary that:

because organizations are open, adaptive systems, we must consider influences and processes that go beyond the boundaries of the company. Our notion of a regularity gradient is intended to span the distance from an industrialized society's global dependence on the continuous and regular provision of fossil energy sources to the local adaptations and struggles and frustrations to deal with uncertainties at a helicopter base in Brønnøysund. Influencing global regularity requirements for fossil fuels will be out of reach for most of us. This does not mean that there are no practical issues to engage in one's own work environment. (Wackers and Kørte 2003)

Recursive Process: Local Optimization

The regularity gradient represents in no sense a determinist relationship causing a large-scale accident. Neither is the transmission of its influence and the resulting induction of vulnerability inevitable. A Norwegian report on vulnerable society (NOU 2000:24, 21), for example, stated that 'to a large extent vulnerability is self-inflicted. It is possible to influence vulnerability, to limit or reduce it'. However, when it is transmitted to local workplaces where people work with safety-critical technologies, it is an influence that induces adaptations in local work processes. It is at the intermediate level of local work spaces that the rational design and rule-based 'mode of ordering' collective work processes meet and interfere with the influence of the regularity gradient. It is also here that design as a mode of ordering fails and where another mode of ordering takes priority.

Design fails for a number of reasons. Among those reasons is the staggering amount of regulations that requires a full-time position and specialized expertise. Accessibility is also a problem on many offshore installations, both in the sense of physical availability of texts and of the language used. Often regulations do not make sense to the people who have to work with them. However, the approach suffers from another problem that has to do with what it means that an activity or practice is rule-based. How can we know that somebody has learned a rule and understands it correctly? How can we be sure that a group of people shares the same correct understanding of the same rule? How can we know that they are able to judge correctly whether and which rules apply in a particular situation? Turner

(1994) explored these issues at length and came to the conclusion that a rule-based notion of practices is untenable. He develops a picture:

> in which practices is a word not for some sort of mysterious hidden collective object, but for the individual formations of habit that are the condition for the performances and emulations that make up life. No one is immured by these habits. They are, rather, the stepping stones we use to get from one bit of mastery to another. (Turner 1994, 123)

In other words, it is the reproduction (imitation, repetition and transformation) of already existing patterns that coordinates collective work processes. We do not have to assume that a rule is being followed when a task is successfully performed. The under-determinedness of the process of reproduction by the existing patterns allows for the (often optimizing) adaptations that constitute change over time and – because the actual adaptation does not exhaust all possible adaptations – also history.

In a similar sense Law (1994, 14–15) uses the term 'recursive process' to describe this self-generating and self-transforming character of social processes, in which the social is both medium and outcome. Contrary to design, recursive process as a mode of ordering or coordinating work is not long-distance but very local. This ordering mode is characterized by the fact that individuals and organizations – but also organizational sub-units such as design teams, maintenance departments or drilling crews – are continuously striving to optimize performance in terms of functionality, effort, time and economic pay-off. The development of an effective work process by an individual or group will be a self-organizing, evolutionary process. This is because an optimizing search is the only way in which a choice can be made among the large number of options for action available in a real-life work environment that is never completely determined or constrained (Rasmussen 1990; 1994). According to Rasmussen (1990), development of expert know-how and rules of thumb depends on adaptation governed by subjective and local process criteria as the work space is being explored to identify convenient and reliable cues for action without analytical diagnosis. Professional performance depends on empirical correlation of cues to successful acts. This recursive process orders collective task performance through reproduction and adaptation (in explorative searches) of already existing patterns of professional behaviour. Success in task completion provides immediate rewards, influences the evaluation of the new emergent ways of doing and the likelihood of repetition. It is in this recursive process that the benefits of the adaptations are immediately perceived, whereas their longer-term consequences often are obscure and intangible. Short-term 'survival' will be favoured over longer-term considerations. When resources become scarcer through misalignments and stress, the tendency to rely on local information, local routines and local standards for what is considered to be good work will increase. The recursive work process from which these local routines and standards for good work emerge explains why sub-systems within one company can exhibit differentiated 'safety cultures' (see Frost et al. 1991; and Richter and Koch 2004 for further discussion).

As a mode of ordering, the recursive process has a number of characteristics and requirements:

1. *Repetition.* A one-time execution of a task is not a recursive process. There has to be repetition, an opportunity to do it again.
2. *Reproduction of existing pattern.* This is a form of repetition, not by the same but by another agent. The new agent does not necessarily have to observe the behaviour of the first agent: but they should be able to sense the more or less durable changes in the physical environment produced by the first-time execution of the pattern.
3. *Adaptation.* Behavioural adjustments in the process of reproducing an already existing pattern; either as a reactive response to changes in the environment or as an intrinsically motivated search for optimization.
4. *Rewards.* There should be some kind of reward system; a way of ascertaining whether the adaptation was beneficial, detrimental or neutral in its effect on the overall system performance. Rewards may be very different in kind: short-time survival; reproductive success; satisfaction of desires; fulfilment of requirements; reduction of effort; increase in speed; saving time, money; praise by a respected colleague; an appreciative clap on the shoulder. Eventually, however, in human work systems, it is the successful completion of a task – the achievement of 'performative closure' – that provides the final reward, influencing the evaluation of all that went into the process of achieving that result.
5. *Memory.* There should be some kind of memory. The memory may be of durable changes in the physical world, for as long as they last. Or it may be embodied in human beings, remembering and drawing on positively rewarded, past experiences. Or it may be memory of a pattern of social relationships that is strengthened by successful collective experiences and in which the threshold for collaboration even in new tasks has been lowered.

Conclusion

It is important to recognize that these two modes of ordering and coordinating a collective work process – design and recursive process – co-exist simultaneously; the primary process of work cannot be reduced to either the one or the other, but interacts in intricate ways, giving rise to robustness and resilience in some and to vulnerabilities in other situations (see Chapter 2 for further discussion of safety-critical systems). As Mol and Law (2002, 11) put it:

> [I]t is not so much a matter of living in a single mode of ordering or 'choosing' between them ... we find ourselves at places where these modes join together

... somewhere in the interference something crucial happens ... complexity is created, emerging where modes of ordering come together and add up comfortably or in tension, or both.

Recursive processes are crucial in overcoming or repairing system imperfections, the kind of frictions that always occur as a result of adaptations and changes in different parts of a system or in its environment. And it is because of local recursive processes that complex technological systems perform reliably most of the time. However, being local, recursive processes are not very good at maintaining coherence over longer distances of space and time; not unless they are fed back into slower process levels in order to transform 'design'. In terms of the account developed here, this is the agenda: recognizing and transcending the limitations of design, recognizing and exploring the regularity gradient, finding ways to re-couple system integrity concerns with business concerns and to protect safety-critical work processes from the influence of the regularity gradient and, finally, finding ways to produce coherence between design and recursive processes.

Postscript on Snorre A

The details of the loss of control leading to the gas blow-out on Snorre A, on 28 November 2004, also entail very local, recursive optimizations in the final stages of planning of a 'slot recovery operation' on well P31A. But these last-minute changes were not coordinated in terms of their consequences for the systems' ability to maintain a sufficient barrier status and physically contain the flow of gas from the reservoir. In the accident investigation the lack of competence could be easily measured against the relevant industry standard (NORSOK standard 2004) on 'well integrity in well and drilling operations', a newly revised version of which had been published on 4 August 2004. However, the analysis of the apparent reduced ability to maintain performative closure starts in 1999. As a result of historically low oil prices (less than $10 per barrel) Saga Petroleum ran into profound financial difficulties. In short, Norsk Hydro and Statoil issued a successful joint bid on Saga Petroleum shares, bought the company together in 1999 and split its assets between them. Statoil acquired 25 per cent of Saga's field shares, boosting its proven and probable oil reserves by about 11 per cent. For Hydro, the acquisition of Saga increased the company's oil production by 45 per cent and its proven reserves by about 40 per cent (Lie 2005, 410 ff.). One of the puzzling elements of the deal was that Norsk Hydro would be the operating company on Snorre A for only 3.5 years, after which Statoil would take over. It was no secret that by taking over Snorre A, Hydro wanted to achieve a substantial cost reduction. After the transfer of operating responsibility on Snorre A to Statoil in 2003, Statoil would be the operator on all fields in the Tampen area, comprising nine platforms, 300 platform wells and 130 sub-sea wells. Because of the geographical proximity of installations that make up a 'result unit' like the

Tampen area it is possible to integrate their operation, logistics and supply into a single cluster and achieve further cost reductions through rationalizations. This was an important business incentive for Statoil to claim operator responsibility for Snorre A in the deal that concluded the joint acquisition of Saga in 1999. From the vulnerability perspective developed in this chapter, however, the quick shifts in operator responsibility on Snorre A are at least questionable. The assignment of operator responsibility on a platform is, or should be, part of the slow processes that provide stability. It takes time to build up personal relationships, to develop specific expertise in a wide variety of technical disciplines. Fast and frequent shifts involving these underlying processes that live on a longer timescale will upset many of the resource relationships a production unit has with other parts of the company. Violating the temporal logic of these nested processes they induce vulnerabilities within the system (local-level imaginative deficit). This will happen even if continuity of local familiarity and experience with a platform is guaranteed through the transfer of personnel from one operating company to another: from Saga to Norsk Hydro to Statoil within a time span of four years. This process was compounded by the negotiation of a new contract with a new 'well operations and drilling contractor' that would take over just a few weeks before the slot recovery operation would commence. Judging from the fact that these changes in drilling contractor and employer, on top of the rapid shifts of operator responsibility, had no consequences for the planning process of a major modification on well P31A, it seems fair to say that the consequences for continued reliable performance have not been explored properly (corporate-level imaginative deficit). The entrainment of safety-critical work in a culture of business performance optimizations created the vulnerabilities that led to a near-disastrous sub-sea blowout (Wackers and Coeckelbergh 2008).

Acknowledgements

I would like to thank the Center for Technology, Innovation and Culture at the University of Oslo for its financial and practical support of my research, and Rogaland Research in Stavanger and the Center for Technology and Society at the Norwegian University for Science and Technology in Trondheim for offering office space during a number of research visits, allowing me to do the empirical research for the case studies.

References

Adamski, A. and Westrum, R. (2003), 'Requisite Imagination. The Fine Art of Anticipating What Might Go Wrong', in Hollnagel, E. (ed.) (2003), *Handbook of Cognitive Task Design* (Mahwah, NJ: Lawrence Erlbaum Associates).

Ancona, D. and Chong, C.L. (1996), 'Entrainment: Pace, Cycle, and Rhythm in Organizational Behavior', *Research in Organizational Behavior* 18: 251–84.
Callon, M. (1998), 'An Essay on Framing and Overflowing', in Callon, M. (ed.) (1998), *The Law of the Markets* (Oxford: Blackwell).
Coeckelbergh, M. and Wackers, G. (2007), 'Imagination, Distributed Responsibility and Vulnerable Technological Systems: The Case of Snorre A', *Science and Engineering Ethics* 13: 235–48.
Frost, P.J. et al. (eds.) (1991), *Reframing Organizational Culture* (London: SAGE Publications).
Goffman, E. (1974), *Frame Analysis: An Essay on the Organization of Experience* (Cambridge, MA: Harvard University Press).
Gunderson, L.H. and Holling, C.S. (eds.) (2002), *Panarchy: Understanding Transformations in Human and Natural Systems* (Washington, DC: Island Press).
Johnson, S. (2001), *Emergence: The Connected Lives of Ants, Brains, Cities, and Software* (New York: Scribner).
Kauffman, S. (2000), *Investigations* (Oxford: Oxford University Press).
Lakoff, G. and Johnson, M. (1999), *Philosophy in the Flesh: The Embodied Mind and its Challenge to Western Thought* (New York: Basic Books).
Law, J. (1994), *Organizing Modernity* (Oxford: Blackwell).
Lie, E. (2005), *Oljerikdommer og Internasjonal Ekspansjon. Hydro 1977–2005. Bind 3 av Hydros Historie 1905–2005* (Oslo: Pax Forlag).
Marion, R. (1999), *The Edge of Organization: Chaos and Complexity Theories of Formal Social Systems* (Thousand Oaks: SAGE Publications).
Mol, A. and Law, J. (eds.) (2002), *Complexities: Social Studies of Knowledge Practices* (Durham and London: Duke University Press).
Mumford, L. (1961), *The City in History: Its Origins, its Transformations and its Prospects* (New York and London: Harcourt, Brace and Jovanovich).
Nicolis, G. and Prigogine, I. (1989), *Exploring Complexity: An Introduction* (New York: W.H. Freeman and Company).
NORSOK Standard (2004), *D-010 Well integrity in Drilling and Well Operations* (Lysaker: Standards Norway).
NOU (2000), *Et Sårbart Samfunn. Utfordringer for Sikkerhets – Og Beredskapsarbeidet I Samfunnet*. Norges Offentlige Utredninger 2000:24 (Oslo: Statens Forvaltningstjeneste Informajsonsforvaltning).
Østebø, R. and Grødem, B. (1998), 'Managing the Regularity Expenditures (REGEX) in the Oil and Gas Industry', in Lydersen, S. et al. (eds.) (1998), *Safety and Reliability: Proceedings of the European Conference on Safety and Reliablility, ESREL '98, Trondheim, Norway, 16–19 June* (Rotterdam: A.A. Balkema).
Rasmussen, J. (1990), 'Learning from Experience? How? Some Research Issues in Industrial Risk Management', in Leplat, J. and de Terssac, G. (eds.) (1990), *Les facteurs humains de la fiabilite dans les systemes complexes* (Toulouse: Octares Entreprise).

Rasmussen, J. (1994), 'Risk Management, Adaptation and Design for Safety', in Brehmer, B. and Sahlin, N.-E. (eds.). (1994), *Future Risks and Risk Management* (Dordrecht: Kluwer Academic).

Reason, J. (1997), *Managing the Risks of Organizational Accidents* (Aldershot: Ashgate).

Richter, A. and Koch, C. (2004), 'Integration, Differentiation and Ambiguity in Safety Cultures', *Safety Science* 42: 703–722.

Roberts, P. (2004), *The End of Oil: The Decline of the Petroleum Economy and the Rise of a New Energy Order* (London: Bloombury).

Rosness, R. (2002), *Feiltoleranse, Barrierer og Sårbarhet* (Trondheim: SINTEF).

Turner, S. (1994), *The Social Theory of Practices: Tradition, Tacit Knowledge and Presuppositions* (Cambridge: Polity Press).

Upton, D. (1996), *Waves of Fortune: The Past, Present and Future of the United Kingdom Offshore Oil and Gas Industries* (Chichester: John Wiley & Sons).

Wackers, G. (2002), 'Vulnerability Profiling: Searching for Sources of Vulnerability in Complex Technological Systems', *Paper presented at the ISCRAT 2002 conference, Amsterdam, 18–22 June 2002*.

Wackers, G. (2004), *Resonating Cultures: Engineering Optimization in the Design and Failure of the (1991) Loss of the Sleipner A GBS* Research Report no. 32/2004 (Oslo: Unipub Forlag/Center for Technology, Innovation and Culture, University of Oslo).

Wackers, G. (2006), *Vulnerability and Robustness in a Complex Technological System: Loss of Control and Recovery in the 2004 Snorre A Gas Blow-Out* (Oslo: Unipub).

Wackers, G. and Kørte, J. (2003), 'Drift and Vulnerability in a Complex Technical System: Reliability of Condition Monitoring Systems in North Sea Offshore Helicopter Transport', *International Journal of Engineering Education* 19(1): 192–205.

Wackers, G. and Coeckelbergh, M. (2008), 'Vulnerability and Imagination in the Snorre A Gas Blowout and Recovery', *World Oil* 229(1): 1–6.

Yergin, D. (1991), *The Prize: The Epic Quest for Oil, Money and Power* (London: Simon & Schuster).

PART II

Accomplishing Reliability within Fallible Systems

Christine Owen

Introduction

Part II of our study of work within fallible systems comprises two contributions that analyze the adaptive strategies (see also Part I), improvisations and negotiations undertaken by actors to reveal how reliable performance is accomplished most of the time despite system vulnerabilities: disturbances, high degrees of uncertainty and unpredictable events. The cases presented in this part include decision-makers directly involved in a neonatal intensive care unit (Chapter 5) and an emergency communication centre (Chapter 6).

As do some of the other chapters in the collection, the chapters in this part focus on what operators on the floor really do. They provide a 'fine-grained', that is, moment-by-moment analysis, the objective of which is to identify how safety is achieved in risky work environments in ways that are frequently unrecognized or misunderstood without that analysis. Although the methodological approaches described in this section share similar features, the conceptual frameworks used in the analyses differ. Messman's study of neonatal intensive care (Chapter 5) draws on concepts from the field of science and technology studies, while Marc and Rogalski (Chapter 6) draw on approaches found within French-speaking ergonomic tradition to develop new ways of re-mediating and thinking about individual and collective operations in mitigating errors and in enhancing safety.

In placing these two chapters together, we draw attention to four ideas either elaborated by or underpinning the arguments of the contributors that signal the movement toward acknowledging the positive role humans play in maintaining resilience and reliability within fallible systems. The ways in which resilience is created and maintained is also discussed in Norris and Nuutinen (Chapter 2); for system vulnerability see Wackers (Chapter 4).

Situating Activity between Procedures and Action

The contributions here demonstrate that it is not the procedures, or the designs of artefacts or work organization that create safety but the work practices of the individuals and collectives involved.

As discussed in the introduction, there is a tension between how much formalized procedure can be provided to govern action in complex environments and how much space is needed for successful action to be built or improvised by operators working in these environments (for a discussion on tensions developed through contradictory systemic development, see also Chapter 2). This tension in work organization has been around for a long time and was at the heart of the work of Frederick Henry Taylor and his scientific management principles developed 100 years ago. The essential characteristic of this method is that work is planned in 'the office', by managers and highly qualified staff, while the production operations themselves are carried out in 'the works' by means of instructions or procedure. Despite changes in work organization since Taylor, there is still often a strict division of labour, vertically between grades of staff, and horizontally, between departments.

The advance in theoretical development evident in the work activities analyzed here is the way they illustrate how Acting Subjects build an understanding of the event and through the process of the event unfolding, manage, manipulate and employ procedures and artefacts within the formal organization of the work to achieve outcomes of efficient and professional action.

The contributions here provide support for a range of emerging empirical and theoretical arguments that full anticipation of action is impossible (Suchman 1987). In work, the operator encounters unforeseen situations and oppositions linked to industrial variability and contingencies (for example, systematic deregulation of tools, instability of the matter to be transformed, etc.). Suchman (1987) used the term 'situated action' to grasp this aspect. Whatever the effort put into planning, performance of the action cannot be the mere execution of a plan fully anticipating the action. This does not mean that procedures or plans are useless, or that guidelines are without interest. They 'guide' and 'help to find the best positioning' (Suchman 1987). However, one must adjust to the circumstances and address situation contingencies, for instance, by acting at the right time and by seizing favourable opportunities. So the message from the contributors here is that there is more to work practice than strict adherence to guidelines. This approach is in opposition to scientific work management approaches, which put forward the hypothesis that procedures can be given because operators work in stabilized working environments. That is, it is at odds with the idea that safety can be regulated through the drafting and issuing of formal safety regulations and the enforcement of compliance with them.

In this section, Mesman's analysis of neonatal intensive care (Chapter 5) focuses on the tensions that are intrinsic in the ways in which medical standards and regulations about newborns are reshaped and how medical staff are 'generally fully aware that complete control of a treatment's unfolding is a fiction'. An inherent part of the work of staff in a neonatal intensive care unit (NICU) is sense-making, and in particular about the risks that are taken. Risk-taking occurs all the time and is often about weighing one risk against another. Mesman argues that in order to undertake successful work in an environment where risk is embedded in almost every aspect of work practice, it is the overall awareness and professional attitude of staff that enables successful accomplishment and that neither regulations nor norms will ever entirely prevent vulnerabilities, near-misses or outright errors. What is needed, according to Mesman, is space for adjustments based on what goes on in actual practices. This is a central question for many of the contributors in the book: What can be anticipated sufficiently well enough to be built into regulations and what must be left to improvised practice in the situation? Mesman shows us this dilemma in the way she analyses signification in practice. In each situation, it is the singularity of care for a particular child and their trajectories in the NICU that are important. Each child, each situation, is unique.

Enhancing Plasticity in Systems

The authors included in Part II enhance our understanding of what might be needed to enhance plasticity within fallible systems. By 'plasticity' we mean the affordances (or capacities) provided in artefacts, divisions of labour, flexibility of rules and procedures that allow the operator room to manoeuvre. Plasticity enables operators to adapt and negotiate their own individual or collective courses of action, to reshape courses of action and to draw on resources as needed. (For another discussion about plasticity see Chapter 7.)

But what are the properties of systems that enable plasticity in complex environments? How do artefacts support systems that provide plasticity for operators? Robinson (1993) describes how what he calls 'common artefacts' (artefacts that can be shared and employed for different uses/meanings) can underpin plasticity and assist operators to maintain:

> an evolving set of rules, understandings, and expectations about the meanings of actions, signs and changes in the common artefact [and when] the participant can also communicate directly using the fullness of their natural language to interpret the concrete situation in front of them. (Robinson 1993, 195)

Robinson goes on to explain that what is needed in systems that support cooperative work where there is a likelihood of unanticipated events are artefacts that provide for shared meaning between operators, without restricting courses of action available, and assist in sharing understanding of the situation through multi-dimensionality and 'overview' components. It is the common artefact that helps workers to build collective sense-making and thus to enable the emergence of plasticity. But the contributors here say more than this: they go beyond Robinson and other accounts of building plasticity into systems. Marc and Rogalski (Chapter 6) show how plasticity is developed and maintained in the collective.

Redundancy is a key concept in enabling plasticity to be available in systems, to ensure workers operating in complex environments are able to address what Rochlin (1999) calls the need for a 'continuous expectation of surprise'. The capacity to respond to unanticipated surprises is critical in complex organizations with high consequences of failure. In organizations, economic imperatives can put at risk levels of redundancy, but as Marc and Rogalski show, this is an important component in allowing the collective to have a high degree of safety, despite the level of errors and mistakes, because of the cross-checking and modification that can be done.

We contend that building plasticity into systems is a key research agenda for the future. The contributions here go some way to articulating what is as yet an under-researched area.

Temporality and Trajectories in Negotiating Action

A common theme throughout the book is an analysis of the temporal dynamics of work in complex systems. The contributions in this Part analyze the role of time in action, but extend this beyond simply analysing the way time is coordinated in contexts where it is a scarce resource (Bardram 2000) or an important factor in work intensification (Gleick 1999). The contributions here analyze the way timeframes can expand or shrink, can run out of phase and can be coordinated and integrated. The relation between how action can disrupt and implode timeframes, or be managed to create 'temporal niches' (to create space for decision-making) is connected to the plasticity available in fallible systems.

The contributors in this Part use the device of 'trajectories' in time to analyze activities and instances of risk, near-misses, critical incidents or accidents. The notion of trajectories is also discussed by contributors in Part I, though in that Part the trajectories that are referred to are broader, longer eras of organizational or institutional development, rather than those discussed in this Part, about momentary situated action. The dynamics of risky work discussed here emphasize the transitional moments – those when the trajectory can go either way, requiring different sets of decisions and actions and resources to be brought to bear. Under these circumstances the plasticity available within the system becomes an important factor.

Interdependencies between Collectives and Human–machine Relations

Messman's and Marc and Rogalski's work adds to the growing body of literature demonstrating that knowledge comes from individuals as well as collectives and their interpretation of their artefacts (Resnick 1993; Suchman 1996; Weick 2001).

Drawing on their respective conceptual tools, the authors express concepts for thinking beyond individual actors and action, such as Marc and Rogalski's 'virtual operator' in emergency centre call-taking and response.

In doing so Marc and Rogalski provide some fruitful new pathways for further investigation as well as identifying aspects not previously articulated. For example, how a nuanced understanding of a collective weighs up the differential effects of possible errors in the course of their activity.

Messman also provides insights into what happens when knowledge needs to be created at the boundaries between professional groups. She analyses the 'competing knowledge frames' that must be negotiated between specialists, doctors and nurses in neonatal intensive care.

Each chapter provides an analysis of the heterogeneous processes that are engaged to successfully accomplish work in the face of uncertainty and doubt, and in working with constraints and contingencies. Thus, the discussions in this Part help us to understand how it is that reliability is successfully achieved almost all of the time, despite the fallibility of systems.

References

Bardram, J.E. (2000), 'Temporal Coordination', *Computer Supported Cooperative Work* 9: 157–87.

Gleick, J. (1999), *Faster: The Acceleration of Just About Everything* (London: Little, Brown and Company).

Resnick, L. (1993), 'Shared Cognition: Thinking as Social Practice', in Resnick, L. et al. (eds.) (1993), *Perspectives on Socially Shared Cognition* (Washington, DC: American Psychological Association).

Robinson, M. (1993), 'Design for Unanticipated Use ...', *Proceedings of the Third European Conference on Computer-Supported Cooperative Work, ECSCW'93*.

Rochlin, G.I. (1999), 'Safe Operation as a Social Construct', *Ergonomics* 42(11): 1549–60.

Suchman, L.A. (1987), *Plans and Situated Actions: The Problem of Human–Machine Communication* (New York: Cambridge University Press).

Suchman, L. (1996), 'Constituting Shared Workspaces', in Engestrom, Y. and Middleton, D. (eds.) (1996), *Cognition and Communication at Work* (New York: Cambridge University Press).

Weick, K.E. (2001), *Making Sense of the Organization* (Oxford: Blackwell).

Chapter 5

Channelling Erratic Flows of Action: Life in the Neonatal Intensive Care Unit

Jessica Mesman

The medical–technological advances of the past decades have been such that entirely new and unprecedented opportunities for treatment have become available. At all levels of the medical domain, new professional routines and sophisticated technologies have radically altered the nature of medical practice. The intervention process has become much more complicated, involving more options, other risks, new decision moments, and more pronounced dilemmas for everyone involved. The ensuing uncertainties have contributed to the emergence of a grey area; one where the established protocols and the conventional answers found in medical handbooks no longer apply. Yet, as before, many clinical pictures still require immediate medical attention and intervention. To ensure a prompt and accurate response by medical professionals, it will always be important for them to rely on protocols and pragmatic guidelines, the effectiveness of which is always subject to enhancement. In addition, it seems increasingly relevant to examine what actually takes place in medical intervention.

The neonatal intensive care unit (NICU), in particular, constitutes a domain where the challenges and opportunities of new medical knowledge and technology converge. As an outpost of today's health care system where the pioneering sprit of medicine reigns supreme, it serves as an exemplary case for studying some of the concrete vulnerabilities in the health system triggered by this permanent dynamic of change. The NICU specializes in the care and treatment of newborns. Very young babies end up in the NICU because their lives are seriously at risk on account of their prematurity, complications at birth, congenital diseases, or potentially lethal infections. This practice is determined by the ongoing flow of activities associated with highly specialized care provision, the admission of new patients (including their parents), as well as by the fluctuations in the conditions of the patients. In this regard, the NICU can be considered a 'High-3 work environment'. See Chapter 8 for a detailed description of the nature of High-3 practices.

This variability requires around-the-clock monitoring and frequent adjustments, while the recovery of an admitted child can never be taken for granted. Life in the NICU is characterized by continuous struggle and entirely unpredictable changes that result in erratic flows of action. A neonate's treatment is like a journey that from the very start is full of unexpected incidents and incalculable uncertainties: both the young patient's destination and the unfolding of its immediate future are all but predictable. Even if the staff has accompanied children on similar trajectories numerous times already, the routes and circumstances are never entirely the same. To some extent, the NICU staff always finds itself in uncharted territory, because a specific technology is new or a certain intervention poses unknown risks, especially given the extreme fragility of the patients. Treatment, therefore, is never merely a matter of solving an infant's medical problem; instead, it always involves the opening up of a very specific, individualized trajectory. This trajectory is characterized by an intrinsic tension between the need for intervention and making adjustments on the one hand and the associated risks and uncertainties on the other (for alternative discussions of work tensions and trajectories see Chapters 2 and 6).

The NICU staff tries to reduce this tension through the application of regulations and norms. But these strategies are not sufficient and are in some cases even a source of uncertainty. My analysis will concentrate on medical protocols as forms of regulation as well as on quantitative data in relation to standard norms. Decision techniques, like protocols, are always open to debate (see, for example, Berg 1997a; 1998). An analysis of concrete vulnerabilities of medical practice requires a focus on 'protocols-in-action' (see, for example, Berg 1997a). Used as basic resources for minimizing uncertainty, as well as the number of risks and mistakes, protocols also function as potential sources for additional risks, errors or near-misses. The flipside of a protocol is that it may become part of the very problem. This same argument can be made for numerical information. Analysis need not be limited to the application of protocols and quantitative data, but can also address their built-in assumptions about the NICU patient and the organization of medical work. It is in these assumptions that the vulnerability of practice itself can be found. Revealing them will not only give us a better understanding of the way in which regulations and norms, initially intended to reduce the intrinsic tension of the trajectory, can act as potential causes of the disruption of the practical order but also of the staff's ability to cope with these conflicting constraints. To reveal the positive role doctors and nurses play in a risky work environment like a NICU, this analysis employs a fine-grained moment-by-moment approach which gives attention to details in the cases and work practice that are analyzed.

Theoretical Resources

My analytical focus is based on the theoretical perspective developed in constructivist science and technology studies (STS). This perspective allows me to criticize the assumption that the fundamental structure of (medical) practice is constituted by principles, deductive patterns of reasoning and decision protocols. Numerous empirical studies on medical work found no evidence for these assumptions. On the contrary, detailed studies of medicine-in-action show how day-to-day operations can never be reduced to a matter of mere application of rules and theoretical principles (see Berg 1997a; 1997b; Lock and Gordon 1988;

Lock et al. 2000; Franklin and Roberts 2006; Rapp 2000, for examples of medicine in the making).

Knowledge is not waiting out there to become applied in practice, but is constituted in the very same practice as it is used. Although there has been substantial research into some of the quantitative aspects of contemporary medical practice, there is still little insight into the complicated interplay of problems and dilemmas associated with the actual processes of managing day-to-day operations in a complex technological system such as a NICU. In this turn to practice, clinical work does not have the sole function of being the context in which busy NICU doctors and nurses are moving around. On the contrary, by focusing on practices, the concrete problems of the intrinsic tension of the trajectory become visible. This does not only concern questions on the kinds of activities that are involved or the kinds of knowledge and skills that are applied in the treatment of newborns. It is also relevant to ask in what way medical standards and regulations about newborns are reshaped in the vortex of concrete activities and how the trajectory is moulded by tensions that are intrinsic to it. Given the increasing intricacy of the overall treatment trajectory, it is equally relevant to go beyond the effects of diagnostic outcomes. An area that needs as much exploration is the linked though discrete, process of diagnosis, intervention and prognosis (see also Mesman 2005).

What actually takes place at the interface of diagnosis and prognosis? Although there has been substantial research into some of the qualitative and quantitative aspects of medical practice, there is still little insight into what is actually happening in regard to the actual processes of diagnosis, intervention and prognostication. To answer this question the focus will be on 'the analysis of intersections', such as lines of work; and of tools, bodies and entities. Hence, this chapter aims to present an in-depth investigation of these processes, where a fine-grained moment-by-moment analysis gives attention to detail in the cases and work practice that is analyzed. Instead of looking 'downstream' at the implications of medical intervention, this study prefers to look 'upstream' into the medical practices.[1] In other words, a turn to practice capitalizes on a careful analysis of the positive role doctors and nurses play in establishing and maintaining a successful treatment trajectory. To accomplish this task, knowledge and practices have to be developed and expanded by the staff members themselves. Like the other chapters in this second part of this volume, this study aims to highlight the individual, collective and systemic competences that are used to enhance the resilience in acting within a complex work environment like a critical care unit such as the NICU.

---

1 Examples of 'downstream' studies, that is, with a focus on the social implications of new medical technologies, are Brodwin (2000); Brown and Webster (2004); Franklin and Lock (2003); while Lock, Young and Cambrosio (2000); Franklin and Roberts (2006) are examples of upstream studies that employ a fine-grained moment-by-moment approach and give attention to details in the biomedical work practice that is analyzed.

I will use two case studies, based on my study on uncertainty and doubt in neonatal intensive care practice (Mesman 2002). My main case involves Robert, a full term baby with a severe malformation of the heart. The complex trajectory of his care and treatment constitutes the core of my analysis. The admission of Peter, a premature baby born after 25 weeks' gestation, will serve as analytical ground as well. The findings of both case studies are based on ethnographic research in two neonatal intensive care units: one in the Netherlands and one in the United States. First, I describe how the staff creates order after a new admission has arrived. The fragile condition of NICU babies requires immediate action. Protocols and quantitative data play a major role in the constitution of a practical order. However, their capability to create or maintain a practical order turns out to be only effective under certain conditions. Second, by following the trajectory of Robert, I discuss how the staff has to work in the absence of two of these conditions: the availability of time and information. Next, I consider the risks implicated in the treatment's transitional moments and explore the organizational level of the work itself. Finally, I focus on the effects that the distribution of knowledge and experience over time and place has on the reliability of the staff's performance.

Creating Pools of Order

Academic Medical Centre

> Bleeping monitors, toiling ventilators, and alarm signals all around. A door opens. An incubator rocks over the threshold, flanked by a nurse and a resident. In the incubator – between lines and tubes – lies Peter, born prematurely at 25 weeks' gestation and weighing just over one pound (610 grams).
>
> Upon his entry, various staff members take action immediately. A nurse places Peter in his permanent incubator and connects the tube in his throat to the ventilator next to his bed. The resident discusses the position of the ventilator with the respiratory therapist: 'Let's put him on 20 to 4, with a frequency of 55 to 100 per cent. If he's doing well enough, we can wean him down.' While the respiratory therapist installs the ventilator, a nurse attaches electrodes to Peter's chest, connecting him to the monitor. A few seconds later a list of numbers appears on the screen: 155, 32, 71, 89, 36.4, providing the staff with information about heart rate, blood pressure, respiratory rate, saturation, and temperature. Another nurse makes the preparations necessary for the blood tests to be taken. Meanwhile, the resident inserts an intravenous line through which Peter can receive fluids and medicine. When finished, the resident and the nurse discuss the medication. Again a list of quantities is summed up.

The attending neonatologist comes in, glances at Peter, looks at the monitor, and back to Peter again. 'How's he doing?' he asks the resident. 'Considering his age, he is doing fine. His apgar-score is hopeful: 3, 6 and 7. I ordered an X-ray, so within a moment we can have a good look at his lungs and check if the tube is on the right spot.' They turn around and walk to the desk while a nurse makes sure Peter is comfortable. He looks exhausted and the monitor displays a fickle picture.

This is how Peter starts his life in a dynamic hospital context. A visitor's first impression of the NICU is one of busy people and noisy machinery. One sees people engrossed in their work, flickering monitors, and extremely small babies, while others – by comparison – look astoundingly big. Sounds pervade the space: alarm signals go off, ventilators are puffing, telephones are beeping, and nurses and doctors exchange entirely unfamiliar words. In this setting the events discussed in this chapter take place.

The admission of a baby in the NICU marks the beginning of a flow of activities and treatment sequences, aimed at answering the question: who is this newly admitted child? All activity is geared towards defining the baby's situation as swiftly as possible, thus inserting him into the unit's daily order. The child's transformation from 'undefined' to 'defined' hardly involves a transparent process, though. Peter's admission reveals the amount of work the staff has to do in order to learn more about who the child is. Such activities constitute a major part of the unit's work; much of the action is geared toward finding out more about the child's condition. The significance of recognizing the source of the baby's problem is not so much in defining its clinical picture, but in the perspective on treatment that it implies (Strauss et al. 1985). Based on the diagnosis, the staff tries to become aware of its implications for the subsequent course of action, including the specific interventions that have to be carried out, their proper order, the tools and the expertise that have to be deployed, and the potential (negative) effects of the interventions. In so doing, the staff relies on established routes that have proved successful: treatment protocols.[2]

---

2 Protocols as being evidence-based guidelines are a clear example of evidence-based medicine as the prevailing mindset of today's health care practice. The drive to improve the quality of health care through standardization and rationalization is not new. From the 1970s onwards we can hear the call for effectiveness, timeliness and patient-centredness. Likewise, the critique on evidence-based medicine is just as old. In de-bunking the rationalization efforts as another technological fix, critics – many from the field of science, technology and medicine studies – have shown how health care systems are simply too complex and too dynamic to be standardized as envisioned by the protagonists. Medicine is not a uniform practice with one kind of knowledge (scientific) which is located in the head of the doctor. A host of in-depth analyses of medical work have shown how messy medical work actually is, and how medical knowledge is socially and materially embedded in routines, its social order, its paper forms, and so on. According to these critics, standardization will not lead to quality improvement. On the contrary, most of them feel that a de-humanization of

The reliance on protocols is a major strategy for coping with risks and uncertainty. A protocol is an ensemble of guidelines that tells physicians and nurses exactly what they should do in a given situation. When planning a treatment trajectory, the staff can make use of these itineraries that guide them through medical situations step by step. The protocol prescribes which diagnostic tests have to be performed, which treatment criteria are valid, in which direction the treatment should go after a certain outcome, when informed consent is needed, or which lab tests are of crucial importance and which ones may no longer be relevant. Based on a conditional if/then pattern the protocol pilots residents and nurses along the planned treatment trajectory. In this way protocols structure the flow of activities in the NICU.

Protocols provide the staff with support and direction in moments of risk and uncertainty. However, they offer no guarantee whatsoever that a child's treatment will actually continue to follow the anticipated trajectory. The route indicated by a protocol is the most ideal trajectory – one with a progressive line. This also requires an ideal patient – one that responds to the treatment quickly and effortlessly.[3] In a NICU, however, one will rarely find such ideal patients because those who are well enough tend to be transferred to high care. The NICU caters to children whose clinical picture is erratic and complex. Their development is rather marked by ups and downs than by straightforward progress. The use of guidelines, which presupposes a more or less linear development, seems at odds with the erratic developmental pattern of children in the NICU. The vital functions of a premature infant like Peter are quite unstable and insufficiently developed. As such, he lacks the capability to self-regulate his body systems properly. His lungs, for example, are too immature to provide for the full amount of oxygen required for survival. In these cases, withholding treatment means instantaneous death. If the staff does opt for intervention, they will immediately connect the baby to the ventilator for oxygen supply. However, the use of a ventilator is not without hazard. Prolonged mechanical breathing with high oxygen concentrations or high air pressure can damage the fragile lungs to such an extent that it reduces the intake capacity to an unacceptable level. In these cases it is hard – and sometimes even impossible – to get the infant off the machine again. The ventilator brought into action as a life-sustaining apparatus will not release the baby. The circulatory system also develops problems due to immaturity. The fragility of the blood vessels, in combination with irregular blood pressure, renders the vessels easily ruptured. High blood pressure can lead to severe brain damage. And there are more problems: a delicate fluid

---

health care and a de-skilling of professional workers will be the main effect of all the standardization efforts. Moreover, standardization neglects the 'hidden work' required to achieve an evidence-based medicine (see Wiener 2000). Timmermans and Berg (2003) share this critique but at the same time call for attention to the problems that have initiated the process of standardization in the first place.

3   See also Guillemin and Holmstrom (1986).

balance due to a thin skin can bring a premature infant (a premie) into a state of supercooling and dehydration; kidneys often have trouble functioning; and there are moments in which the premie 'forgets' to breathe because of the immaturity of the respiratory mechanism in the brain. In response to this, the heart rate can drop rapidly.

Considering this list of possible dangers premature infants must cope with, Peter's treatment resembles tightrope-walking. A small change in his health status can have dramatic consequences. The staff is very aware of the infants' vulnerability and the unintended consequences of their own interventions. This means that the staff attempts to keep close tabs on the vital signs in order to be in the best possible position to prevent disastrous outcomes. Therefore, the most accurate, up-to-the-minute information is required, often on an ongoing basis. The more control, the better are the chances that Peter will survive the incubation period without serious complications. However, the staff is well aware that at any time, without any warning, he can suddenly collapse. Hence the intensive monitoring practices. Every irregularity, no matter how small, activates an alarm attached to the monitor. Every hour a nurse checks the outcomes displayed on the monitor and records them on a list next to her own observations and the results of regular blood tests. At regular intervals, the resident carefully examines the baby as well.

Thursday 9:10

Peter is asleep. The medical form on top of his incubator tells us he was born at 25 weeks of term and his weight is somewhat more than one pound (610 grams). Compare these figures with those for a full-grown baby (40 weeks of term, c. 3000 grams, that is six pounds) and one realizes how small Peter is.

At a desk in the middle of the unit sits a resident. He reads the previous night's report about Peter and makes some notations. Next, he examines the information recorded from the life-support systems and monitors. He studies a long list filled with all kinds of numbers: input and output rates, body temperature, ventilator settings. He calculates means and ratios, intravenous feeding, medication, intake, and the adjustment of the ventilator. He writes his conclusions in the record and compares this information with the results of the laboratory tests.

All of this information, combined, paints a specific picture. With this impression, the resident approaches Peter's incubator for the physical examination. After washing his hands he listens to the infant's heart and lungs, feels the fontanelle,[4] and evaluates the skin colour. After a few minutes he walks back to the desk again and writes down his findings. Now, by arranging and comparing data from different sources he is able to make a determination of Peter's condition upon

---

4   A soft membranous space between the cranial bones of an infant.

which he readjusts treatment. During rounds he will discuss his findings and treatment decisions with his supervisor.

The resident assesses Peter's situation mainly while sitting behind a desk, with pen and calculator in hand. These do not seem to be the proper tools for examining a premature baby, nor would one expect a NICU resident to spend much time behind a desk. But a physician who has to assess a baby's condition does not only look inside the incubator; he also studies the record, the most recent data and the monitor's current data. Various elements converge around the baby: numbers, words, and stacks of paperwork, devices, instruments, bodies, decisions and the architecture of the ward. The table in the middle of the NICU functions as the ordering centre. Such a site is constituted by 'gathering, simplifying, representing, making calculations about, and acting upon the flow of immutable mobiles coming in from and departing for the periphery' (Law 1994, 104; see also Latour 1987, 227 for a discussion). This is where the resident compares and combines the various data, so as to turn them into a coherent, unambiguous whole. The record functions as a central node. The day's report, the test results and the outcome of the physical examination are added to the record. Writing is not so much geared towards communicating information, but towards ordering information (see also Berg 1997a for a discussion). In the record everything comes together and this results in a narrative sequence of the baby's changing situation. The record is like a travelogue that depicts the baby's itinerary during its hospitalization. A single glance at this information directly informs the physician of a child's current situation, including its most recent adventures. In this way the record serves as a centre of representation (Law 1994, 26).

What comes to the fore is how Peter's admission unleashes a flow of quantitative information and activities. The baby is connected to machines with numerical input and output; lab tests are executed, yielding fractional data; nourishment and medication are carefully calculated, calibrated to a specific pace. Staff members collect numbers, calculate ratio and proportions interact in numerical discourse, and list numerical outcomes on special forms suited to that purpose. As a result, the condition of the neonate is increasingly represented by a variety of numbers: numbers related to technological devices as input or output measures; numbers generated by the staff, such as pulse rate or calculated ratios; numbers that are visualized and displayed on a monitor or verbalized in the discussion during rounds. Hence, calculating is a constant, recurring activity and, as such, constitutes a significant part of the NICU's daily routine activities.

Through exquisite technological instruments, all kinds of body processes are measured and visualized in a numerical language. There is a proliferation and ubiquity of quantitative data on the NICU. Because of its factual precision and comparability, a numerical transcription of the body is considered more reliable than its qualitative counterpart. This quality is thought to eradicate the subjective part of knowledge, which is seen as an obvious obstruction in the mutual exchange of data. The unambiguous nature of numbers makes possible a joint discussion of data. Numbers are considered hard facts, objective measurements of reality. It is not subjective estimations, but rather a standardized method of reasoning that reveals in an objective way wherein the problem lies. In the ceaseless collection and comparison of numbers, medical personnel find a compass to guide them.

From the very start Peter's identity is established in a process of reading, measuring, calculating, comparing, combining and discussing data, but also through sensory techniques such as observation, feeling and listening. The definition of the infant's condition is not the outcome of an indiscriminate accumulation of data, but has to be a carefully considered composition. The fluctuating condition of the infant causes the established order to be only a momentary one. In this context, it would be misguided to believe in 'pools of order'; rather, there is an ongoing process of 'ordering'. As Law suggests, 'orders are never complete. Instead they are more or less precarious and partial accomplishments that may be overturned. They are, in short, better seen as verbs rather than nouns' (Law 1994, 1).[5] The medical staff is aware that its tools are limited and that the data they work with tend to have a high turnover rate. Given the fact that new data may undermine the established clinical picture at any moment, physicians rarely aim to determine an infant's true or ultimate identity. After all, this would be useless. This is why they are satisfied if their grasp of the infant's condition provides sufficient information to allow them to do the next step of the treatment trajectory. Being aware of the high turnover rate of data, the staff intends to establish a practicable identity instead of a true identity.

My analysis of the first few hours of medical work following Peter's admission shows how protocols and numerical data act as standard resources to produce a reliable performance based on complex trajectories. Protocols and numerical data play an important role in charting the course of treatment. A protocol functions as a road map that offers the best route for reaching the final destination and the daily test results indicate whether the infant is still on the right track. By translating physiological processes into a numerical language, detailed measurements become possible. A comparison among past, present and future medical data displays the course of internal body processes through time. Is the infant more or less stable? Is it recovering or is its condition worsening? Through the evaluation of factual representations of various internal physical processes, the medical staff attempts to detect the presence and extent of pathological processes. However, in the description above the situation is clear and the staff knows what to do. Peter is a premature baby and needs to be taken care of according to the standards of practice in cases of prematurity. In other cases, however, the available protocols and quantitative

---

5   Therefore, study of the process of ordering is more appropriate than study of the order itself. Order and ordering processes are major themes in sociology, especially in micro-sociologically oriented directions such as ethnomethodology. Put briefly, ethnomethodology focuses on everyday rules, on the recipes people rely on for organizing their life on a day-to-day basis. Garfinkel (1967) and Lynch, Livingston and Garfinkel (1983) are examples of ethnomethodological studies, the latter being specifically geared towards scientific practices.

data may be less unambiguous. Protocols and quantitative data assume clarity that certainly in the NICU is not always readily available.

Decision-making in a High-density Zone of Doubt

As a set of instructions a protocol tells the staff what to do in specific situations. On the basis of its 'if ... then' structure a protocol guides staff members along complicated treatment trajectories. As Berg points out: 'the protocol functions as a focal point of reference to which different staff members refer, can orient themselves, and can find instructions on what to do next' (Berg 1998, 232). A protocol delineates the roles and tasks of the nurses and physicians performing a coordinating role. To fulfil this task, doctors and nurses have to delegate part of their coordinating activities to the protocol.

To act as a coordinating tool, a protocol assumes a specific practical order, including the availability of diagnostic information or the time to collect this information.[6] But what should be done if both are absent? The need for making specific choices is most urgent right after a baby's birth, when generally there is minimal insight into its condition. Yet such medical knowledge is essential for choosing the protocol to be followed. How should one proceed? Which strategy does the staff deploy to reach reliable performance despite the odds? The admission of baby Robert is used as a case study for developing this concern.

Robert is born in a peripheral hospital. After his birth his health seems perfectly okay, but when the physician cuts the umbilical cord the baby turns blue all over. The physician has no information on what causes it, but he has to act immediately in order to keep the child alive. How to act? How can he select a treatment trajectory in the absence of a diagnosis? The selection of a protocol assumes the availability of a diagnosis. In the absence of a protocol and being fully aware of the seriousness of the situation the physician contacts the Academic Medical Centre's NICU right away.[7] This decision will prove to be Robert's rescue.

Academic Medical Centre

> Robert is taken by babylance (that is, an ambulance specifically for babies) from the peripheral hospital to the NICU. The staff, awaiting his arrival, is busy making preparations. There is an open incubator with all the necessary apparatus, while the pediatric cardiologist has been informed of the infant's impending

---

6 A protocol presupposes the categorization of the child in a specific diagnostic disorder. In *Sorting Things Out*, Bowker and Star (1999) describe how doctors sort perceived characteristics into categories and the consequences of those choices.

7 Although most Dutch hospitals have a medium and high care unit for neonates, only academic hospitals have intensive care units for them as well

arrival. The cardiologist examines Robert immediately. The monitor of the Doppler-machine clearly shows how his oxygen-rich blood directly flows back into the lungs. There is no doubt about the transposition. After a talk between the cardiologist, the neonatologist and the heart surgeon, it is decided that corrective surgery will be performed.

Robert appears to have a transposition. This is a disorder of the heart that causes oxygen-rich blood from the lungs to flow via the heart directly back into the lungs again. This means that only low-oxygen blood enters the circulatory system. This disorder can be corrected by surgery. Without surgery there is no way that the child will survive. But this kind of surgery is risky: it either succeeds or fails; there is no in-between. If it succeeds, there is a maximal return: the child will grow up like any other healthy child.

Diagnostic tests may confirm the cardiologist's suspicions. Doing such tests, however, takes time, while in acute cases like this one immediate action is called for. The physician in charge at Robert's birth has to choose a trajectory on the basis of incomplete knowledge. Obviously, this is not without risks, and therefore the physician does not only select a treatment trajectory, but also – and perhaps more so – the kind of risk he is willing to take.[8] The physician did not first try to establish a diagnosis, but immediately opted for admission to a more specialized department at another hospital. Thus he sought to limit the risk of Robert dying.

Time is a scarcity in the NICU. Frequently there is neither time for extensive diagnostic testing, nor for carefully prepared interventions. For treating Robert's heart disorder a custom-made trajectory is designed involving the surgery, the medication, pre-care measures, the aftercare, the timeframe and the possible complications.[9] Given the complexity of the trajectory that is called for in Robert's case, detailed and time-consuming preparations are absolutely necessary, which means that somehow extra time needs to be created. This is done through medication that slows down the closure of the *ductus arteriosus Botalli*. This ductus is a channel in a fetus's heart that conveys the blood from the aorta back into the lungs. After birth this opening – the ductus – closes within 24 hours, so as to allow the oxygen-rich blood to enter the circulatory system. When the ductus fails to shut, the oxygen-rich blood will flow back into the lungs. This will fill up the lungs, while the blood in the body's circulatory system does not contain enough oxygen. By keeping Robert's ductus open, the oxygen-rich blood may

---

8   If physicians decide to refrain from intervention, they run the risk of letting a child die that still has a chance of survival. Yet intervention may also come with the risk that the child will have a life entirely devoid of quality. On this dilemma, see also Frohock (1986), Guillemin and Holmstrom (1986), Strauss et al. (1985) and Zussman (1992).

9   Although notions like 'protocol', 'standard' and 'guideline' tend to be used interchangeably in the literature, I use 'protocol' to refer to a standardized (and hence commonly shared and broadly supported) form of acting.

still enter the circulatory system. The decision to apply this medication provides the staff additional time – a temporal niche that allows it to make preparations for the next steps of the complex trajectory. One of these preparations involves an extensive clinical lesson for the staff involved.

Monday 14:00

> There is a clinical lesson for the nurses about tomorrow's transposition. The meeting is well attended. With the help of slides the child cardiologist explains the problem and tells them what will happen during tomorrow's operation. Next, he discusses everything that has to be done and what deserves special attention before the child leaves for the OR as well as after its return to the NICU.
>
> The cardiologist emphasizes that a 'heart baby' like this demands another attitude: 'Normally we have to do with insufficient lung function. In those cases there is still a little time to discuss what you are doing. You change the settings of the respirator or things like that. Basically, every move can first be discussed. But a child with this kind of heart problem requires immediate action. There is no time to discuss options in some other room. Therefore it is crucial that in advance you closely study the protocol I have written. Although the protocol is essential, it is the actual situation that at all times will determine what should be done.'

Protocols presuppose a diagnosis, which in turn provides the key for sound decisions on the various treatment options. But in Robert's case there was no time to collect diagnostic data. In such a situation there is a shift from having to choose a specific treatment trajectory to having to choose the kind of risk that is taken. This issue changes from 'what is the preferable treatment trajectory?' to 'what is the least desirable risk?' Furthermore, protocols enforce a timeframe, if only for doing diagnostic tests and making the necessary preparations. Time is obviously a scarcity in many cases, while it may also be a major risk factor. In such cases, staff members will try to create a temporal niche that gives them extra time to collect additional data and prepare the next step as well as possible.

Moments of Transition

Tuesday 10:00

> This morning at eight o'clock Robert entered the OR. There is a tense atmosphere in the NICU. Repeatedly one can hear someone inform how the operation is going and how much longer it will take for Robert to return. The attending neonatologist is somewhat irritated by all this attention: 'So much concern for that one child. All the other children are important as well.'

Obviously Robert's condition is so fragile that the surgery is extremely risky. As a step-by-step planner, a protocol always contains several moments of transition. Frequently, these transitional moments function as moments of risk. During the operation such transitional moments occur from the very beginning. Once Robert is disconnected from the heart–lung machine, for instance, he himself must take over these vital functions, otherwise he will die. The cardiologist is very aware of this particular risk:

> 'These cases cause one a lot of stress. If everything goes well, I feel like conquering the world tomorrow. I really get a kick out of it. But if things go wrong with such a child, I feel terrible for a few days. I am extremely exhausted and unable to handle anything. It demands so much from you. You give all you have in such a case.
>
> There are always a few anxious moments during which you are afraid that things will go wrong. One such moment is when the child is disconnected from the heart–lung machine and he has to perform those functions on his own. If he does not do this, you are stuck. In the previous case it did not work out and the child died after all.'

In Robert's case it did work out well and he is brought back to the NICU. With the successful disconnection from the heart–lung machine a major hurdle in the trajectory is taken. Although the heart surgery is successful, there is no reason yet for celebration. For the first 48 hours after surgery there is still a substantial risk that Robert will die. The next transitional moment presents itself right after his return in the NICU:

Tuesday 17:00

> Robert returns to the NICU, accompanied by a 'green swarm' of cardiologists, anaesthesiologists, neonatologists and nurses. It is hard to see a baby at all, enwrapped as he is in lines and tubes. Taking Robert out of the reanimation cart and putting him into his incubator involves a tricky procedure. Much has been said about this moment during the preparatory session. As foreseen, there is a spaghetti of lines and tubes that can hardly be unravelled. Extreme caution is called for, but it also has to be done fast, for the child is extremely vulnerable. The sooner he is comfortably back in his incubator, the better. After carefully lifting him with all his technological baggage, they prudently put him in his incubator. Everyone is looking tensely at the monitor that briefly shows his levels jump up and down, after which they become stable again. It worked out well and the staff members are all relieved. Yet immediately afterward the neonatologists and cardiologists begin to discuss the proper interpretation of the numbers. Children have other levels than adults, the neonatologist claims. But, as the cardiologist

argues, a cardiac patient has yet again other levels than a normal neonate. This struggle will go on in the next few days.

In order to become effective, quantitative data must be embedded within a frame of reference. Outside of an interpretive framework, numbers lose their meaning. Numbers are not isolated entities; they are related to other numbers within particular contexts of practice and meaning. Where numbers correspond to other numbers, each set may be made more meaningful. But when they contradict each other, this may give rise to doubt and competition over the distribution of doubt and certainty. A closer look reveals that such knowledge frame is no guarantee that medical decisions and interventions will proceed smoothly. For in the case of Robert there is not one but two competing knowledge frames: a neonatological and a cardiologic frame of reference. Because both specialists felt they were right, turning data into medical action required an ongoing process of negotiation.

Tuesday 19:00

Robert's chest is covered with a large band-aid. During the last half hour the band-aid has slightly bulged. All his parameters are okay except his venous saturation level.[10] The staff decides not to wait for the other parameters to go down as well, because the message of the venous saturation is clear: the child suffers from a bleeding. Direct action is called for, leaving no time to bring him to the operating room (OR), which is too risky and not absolutely necessary. While the cardiologist is putting on his sterile garb, two nurses do the required preparations. On a cart a sterile field is created. The surgical tools are put on the lower part of the cart. One staff member goes to the OR to pick up a suction pipe. The anaesthesiologist is ready as well.

The plastic band-aid is cut open and the old blood that has accumulated is being drained off. The larger clots are removed with tweezers and weighed. Probably these blood clots clogged the drain, which no longer allowed the fluid from the wound to be discharged. This caused higher pressure around the heart, which reduced its pumping power. The draining diminishes the pressure and the parameters on the monitor slowly improve. Initially things do not seem as bad, but quite soon fresh red blood appears. The pressure has indeed caused a bleeding. The cardiologist puts on his surgical glasses and starts looking for the leak. Meanwhile Robert receives a blood transfusion. Soon the leak is found and the rupture closed. As the bleeding stops, the cardiologist begins to curse as a way to express his relief. He remained calm throughout, but now the danger is over his stress comes to the fore. He managed to stop the bleeding in time. Everyone responds with relief. The monitor clearly indicates how Robert is coming out of his crisis.

---

10 The venous saturation level indicates the oxygen level of the blood that flows towards the heart.

Although not yet corroborated by related parameters, the outcome of the venous saturation was considered as a fact hard enough to validate an intervention. Different from the monitoring machines, the doctors could *anticipate* what was coming. Machines can only register and react, but not anticipate the near-by future. The doctors' decision to act proves that the interpretive framework of quantitative data is not limited to a list of numerical standards. Numbers are not only corroborative or conflictual with each other, but they are also linked up with other forms and sites of knowledge, such as clinical experience, clinical symptoms, and diagnostic non-numerical representations such as x-ray and ultrasound images or as in this case, a bulging band-aid. For experienced doctors the meaning of quantitative data exceeds labels like 'normal' or 'irregular'. Their interpretation is more a practical matter of what is acceptable or unacceptable. The saturation and the bulging band-aid might be indicators of a bleeding. A change in the other parameters might or might not confirm this interpretation. The risk that then they might be too late to save the child made them decide to act immediately and operate on Robert in the NICU. Deviations from standard procedures are sometimes necessary to prevent dramatic outcomes. In cases like these medical personnel look for a compromise between assessing the child's condition completely and saving time to act. They try to find a 'sufficient' situation assessment for action under uncertain conditions.[11] By deviating from the standard frame of interpretation, as well as from the protocol rule that operations have to be performed in the operating room, the child was saved. This was also possible because the access to the heart was not closed after the operation but merely covered by a patch. By taking into account the possibility of a bleeding in the protocol for Robert, the staff was prepared for it. Once the bleeding indeed occurred, immediate action could be taken: after all, there was still open access to the heart. It was merely a matter of removing the patch that covered the wound.

Risks differ in their level of probability: it is always uncertain if complications will indeed occur. At the NICU, however, risks are considered an inevitable element of everyday practice. One way to deal with risks is to ignore their probability and consider them as established facts. The staff acts as if a specific risk will in fact become a reality, until the evidence is otherwise. Unforeseen and nontransparent situations, in other words, are taken into account in the planning of the treatment trajectory. In the guidelines for Robert's treatment all the possible complications are calculated in advance. That Robert may have a bleeding after surgery is not seen as possibility but as certainty. This in turn generates another choice: the choice between one risk (a bleeding that cannot be reached fast enough) and another risk (an infection). In Robert's case the first option is chosen – one that only a few hours after the surgery proved to be justified.

---

11   Cf. the role of compromise in the model of ecological safety. See Chapter 6.

The keeping open of this 'emergency entrance' ultimately turned out to be lifesaving. But it cannot be kept open forever. The closure of the wound on his chest equally constitutes a risky moment of transition.

Thursday

Robert's condition has noticeably improved. He is given a bath and his skin has more colour. His drain still barely generates fluid. This is a good sign. Now that his condition is stabilizing, the monitoring team is lowered from three physicians and two nurses to two physicians and one nurse. The number of machines to which he is hooked up, however, is not reduced yet.

Friday

Robert's chest will finally be closed. Around his incubator a sterile field is created once again and a team of doctors and nurses prepares for surgery. First the band-aid is removed, followed by the patch that covers his heart. Each stitch is cut carefully. In the chest's cavity the heart, beating calmly, becomes visible. The drain is put to the side, so that the catheters can be removed that measured the pressure in the heart: first the catheter from the left atrium, then the one from the right atrium. At this point something goes wrong. When removing the second catheter something else is pulled loose. The effect is a bleeding. The cardiologist acts instantaneously. 'Where is the extra blood?', the neonatologist asks the nurse at his side. 'It has been ordered already, but it is still at the blood transfusion department.' The neonatologist appears shocked: 'Go get it right away.' The nurse leaves running, while the neonatologist pulls the other nurse aside: 'The blood should have been here. Here in the ward. Not at blood transfusion. The bleeding is not so bad yet. But if it were worse, we would have been in serious trouble.' The nurse fully agrees, and next time it will be done differently.

To everyone's relief, the cardiologist manages to halt the bleeding. Now the chest can be closed. Layer after layer, everything is sewn up. First the breastbone that still only consists of cartilage; next, the dermis that is mainly fat; and finally the outer skin. Each layer has its own knots and stitches. The inner layers are done with dissolvable stitches, unlike the ones of the outer layer that will be removed later on. After everything is sewn up, the nurse washes Robert and this makes him look much better already.

Despite the cardiologist's great care, he could not prevent a bleeding from occurring. This is a risk with which the staff is familiar, which is why extra blood was ordered. This blood, however, had not yet been delivered to the NICU. A cool-headed physician, a nurse in good shape and some luck contribute to averting a bad outcome. In practice errors act not only as cues indicating limits but also as occasions for learning.[12] (See also Chapter 2 for further discussions on learning.) For

---

12   Much research on errors has been done in complex systems, such as the aviation industry and nuclear power plants. Although health care differs in many respects from these

staff members, such near-misses are also learning moments, from which the next child in open heart surgery will benefit. The unfortunate experience with Robert in this respect will challenge the staff to ensure that in the next child's treatment plan the availability of extra blood in the NICU will be listed as a prerequisite. In the treatment trajectory, guidelines may give shape to actual interventions, but these interventions in turn lead to more knowledge and new guidelines.

Distributing Knowledge and Experience

Moments of transition in treatment are moments that carry extra risks, but the same is true in a literal sense regarding the inevitable moments of transition in the organization of daily activities in the NICU. A daily moment of transition, for example, is the staff's rotation, which involves interactions and the transfer of information. One of the specific risks is that data are communicated or understood incorrectly. A complex trajectory like that of Robert requires the input of a variety of medical specialists, including cardiologists, anaesthesiologist and neonatologists. In such intricate cases it has to be clear who has final responsibility, as misunderstandings have to be avoided at all cost. This is why the collaboration between the various medical disciplines and the mutual gearing of the various sorts of medical knowledge are of great importance. In this respect one of the neonatologists of the Academic Medical Centre observed the following:

> 'The child is in our ward, so ultimately we are responsible. With this kind of surgery the cardiologists of course have more experience. But we have more experience with infants. Much deliberation, therefore, is needed, which is not always easy. One interprets the saturation levels of children differently than those of adults, while a cardiac patient has yet again different saturation levels. What we consider too high, a cardiologist may view as too low or the other way around. But it would be wrong if a nurse were told by one specialist to do this while another tells her to do the opposite. There is an agreement that we as neonatologists are in charge here and that therefore we are consulted on everything that takes place. The nurses only do what we tell them to do, so as to avoid conflicts. This does not mean that we tell the cardiologist: "You have to do this and that.' No, it is a form of deliberation in which you propose solutions and subsequently decide things together."

Treatment, then, is not just a matter of simply following a fixed trajectory, but of ongoing discussion and a collaborative effort to find common ground between the child's actual medical development and the projected course that will lead to

---

practices (see, for example, Randell 2003) it can improve their safety by learning from their experience. As found by Tucker and Edmondson, (2003), hospital organizations often do not learn from operational failures.

its recovery. Intensive monitoring of the child allows the staff to notice and correct sudden changes in its condition as rapidly as possible. This task is especially in the hands of the nursing staff, who rely on the constant flow of information on the various bodily functions provided by the clearly visible monitor to which a NICU child is connected. Electrodes that are attached to the child's chest register various activities, such as heartbeat, respiration rate, saturation and blood pressure. If the parameters exceed a specifically indicated margin, an alarm will sound to warn the staff. In addition to this around-the-clock monitoring, every three hours a nurse checks the child's physical condition. If its stability declines, this frequency may be increased accordingly.

Right after Robert's surgery his instability is such that in the first hours his care is entirely a technological affair: he is only touched, so to speak, by the tubes of the many machines to which he is connected. This does not imply that the staff has nothing to do. On the contrary, Robert requires all the attention they can give, but most interventions take place outside the incubator. Caring for him means checking devices, comparing results, writing down levels and adjusting the settings. Rather than being an obstacle, technology functions more or less like a material passage through which the child has to pass before it can be returned into the hands of the nursing staff. The extreme instability of Robert's condition requires the utmost concentration of all staff members involved. His record consists of so many parameters that it is no longer possible to identify intervals between the various checks; those in charge of him hardly pause for a moment. Robert is not just a neonate but also a cardiac patient. For such patients other rules apply and other forms of reporting data are used. Since this kind of record is rarely used in the Neonatology Department, Robert's monitoring is an especially difficult matter. So far this particular NICU rarely admitted infants with transposition, which means that the nursing staff has had little opportunity to gain experience and build routine with this type of case. Therefore, when adding new data to the day's list, closer attention and double-checking are called for.

The complexity of Robert's monitoring proves to require a kind of specialized knowledge and experience that is still hardly there. Although this NICU certainly can supply the expertise needed for his treatment, it turns out to be hard to distribute this expertise in a balanced way among the staff that is present at any given moment.

Thursday 15:00

When the nurses of the evening shift enter, a light panic erupts. No one of the certified nurses feels experienced enough to take over the responsibility for Robert's care. As one of them says: 'These children are always cared for by the same staff members, as a result of which others, including me, do not gain experience in treating them. But now you cannot expect us to do this straight away. Just look at how this child is hooked up to everything. I cannot do it.' To solve the problem, an experienced nurse scheduled to work in the High Care Unit changes place with one of the less experienced NICU nurses.

Is Robert a cardiology patient or a neonatology patient? How do you know where to enter numbers on a form that you have not seen before? Because of the staff's rotation and specific disciplinary differences between neonatology and cardiology, providing intensive care is not only a means to rule out uncertainties, but it also becomes a source of uncertainty itself. The best staff members for special cases cannot be present around the clock, nor does each staff member have multiple years of experience. This not only applies to the nursing staff; doctors, too, do not all have the same knowledge and experience. For example, how, as resident, do you get through your night shift with a child like Robert? What is the proper interpretation of the values on the monitor? Will there be a neonatologist around tonight as well? In a complex case like Robert's, as a rule experienced physicians are present at night and during weekends. And for a good reason.

Friday 22:00

Unexpectedly, the saturation level of Robert's blood drops sharply. The attending nurse just left. The resident is hesitant about what to do. Everyone is anxiously looking at the monitor screen. At that moment the anaesthesiologist and the pediatric cardiologist enter the NICU. They intervene instantly. After some minutes the monitor indicates that the saturation level improves again. The anaesthesiologist is grumbling at everyone because upon entering she noticed that instead of doing something they were all just watching the monitor.

Significantly, then, numbers do not tell the staff members what to do: they have to decide for themselves. In so doing, the numbers may serve as basis. As we have seen, quantitative data, in order to become effective, must be embedded within a specific frame of reference. However, the availability of such a frame does not guarantee immediate action. After all, the resident knew that a decreasing saturation level indicated a serious problem, but nevertheless did not know how to act. Confronted with an irregular outcome he is unable to evaluate, the resident does not know what action to take. Junior residents, especially, lack the necessary experience at moments like these (Anspach 1993; Zussman 1992). During their first semester in the NICU, for example, there are many things they have never seen before. Realizing the gaps in their medical knowledge, they may become very insecure about the accuracy of their interpretation.

Monday

Robert is doing well. His condition improves noticeably. His technological pack has been substantially reduced. If last week there were eight lines and tubes that supplied him with medication, extra blood, oxygen and nourishment and that drained his urine and the fluid from the surgical wound, now only two or three are left.

Robert's treatment demonstrates the potential of neonatology at its extreme. After being a child with little chance of survival he became one with a future like that of any other healthy child. His situation remained stable and his chest could be closed forever. Gradually, he was rid of his technological pack, which allowed the re-emergence of the baby in him. In due time he recovered visibly. After two weeks he did so well that minimal monitoring sufficed. He was returned to the hospital where he was born, from which he was dismissed again after just one week. Sent home, healthy and well.

As we have seen, numbers function much like a compass that guides NICU staff members in their effort to pilot the infant safely along the planned treatment trajectory. By constantly gathering numbers, calculating ratios and quantities, entering quantitative data on special forms and classifying them, and by interacting with each other in numeric language, the staff is able to monitor the condition of the child. Numbers, however, do not speak for themselves, nor can a large flow of quantitative data do away with each and every uncertainty: medical intervention trajectories in the NICU are always rife with risks and uncertainties. Quantitative input and output are tools fabricated through their own applications in practice. This contributes to a tension between contradictory numerical frameworks. On the one hand, numbers are considered as objective reflections of reality; on the other hand, they are also constitutive elements of this same reality.

In other words, numbers do not possess a solidity of their own, but become facts through processes of negotiation. Although claiming and seeming to rely on the exactness and objectivity of digital data, the staff engages in continuous negotiations about their meaning. A closer look at their daily activities reveals the contextual reframing of numbers. They are related to other numbers within particular contexts of practice and meaning. However, detailed analysis reveals that the existence and knowledge of such a framework is no guarantee that medical decisions and interventions will proceed smoothly. The distribution of numeric certainty and credibility appears not to be arbitrary, but is rather affected by contextual constraints and possibilities. Again, there can be a mutual confirmation of numbers or continuing struggles for credibility. Analysis of this contingent interaction shows how the meaning of numbers in terms of true or false, certain or meaningless, is the outcome of a distribution of credibility within a network of quantitative data and other frames of knowledge. Hence the meaning of numbers is shaped by specific contexts.

When a particular trajectory is marked by erratic and unexpected turns, this does not necessarily imply that the staff is unable to bring an infant's treatment to a happy conclusion. Experience teaches them about the risks that turn up at specific moments. These risks are seen as given and taken into account in the treatment trajectory's planning. My account of Robert's treatment brings to light how the staff can pilot an infant along an extremely risky and complex trajectory through preventive measures, constant monitoring and a continuous flow of data on its current condition. Precautionary measures and ongoing monitoring may avoid much suffering, but they can never offer full protection against undesirable developments. Calculated risks or unanticipated complications may still cause a treatment trajectory to founder.

Although uncertainty is a major component of medical knowledge in a NICU, this does not mean that the staff members constantly feel its burden when performing their daily activities. In actual practice, the uncertainty that is always part of medical knowledge is compensated by the suggestion of control, which tends to characterize medical acting. Routines, clinical experience, skills and knowledge – all contribute to a position from which physicians can begin to act in ways in which they are used to acting (see Atkinson 1984 and Katz 1984 for further discussion). Based on common-sense reasoning and a pragmatic attitude they make decisions and perform actions, thus finding their way, generally fully aware that complete control of a treatment's unfolding is a fiction.

Applications for Practitioners

A focus on the complexities of human decision-making in the face of uncertainty allows us to identify the 'hidden competences' that are required to ensure the continuity of medical intervention in the NICU despite high degrees of diagnostic and prognostic uncertainty and unpredictable events (see Wiskerke and van der Ploeg 2004 for a discussion on 'hidden novelties'). Therefore we have to look at the synergy between coordination tools like standard norms and regulations, and the expertise of the staff members. What is the interaction between the expertise of the staff and the capabilities of the health care system in which they operate? The implication of a focus on the synergy between procedural standards and professional work is that it will *not* be about the product of action or, for that matter, about more or less experience and/or expertise. Instead the focus needs to be on: processes of alignment, fine-tuning and synergy between the involved forms of knowledge in the constitution and preservation of a successful treatment trajectory. This type of research can be best considered as an attempt to 'exnovate': to explicate the competence and expertise which is available in the specific practice and of which the actors involved are not aware themselves![13] Importantly, more than innovation, exnovation does justice to the creativity and experience of those involved, in their effort to assert themselves in the particular dynamic of the NICU practice. Exnovative research reveals hidden processes and competences involved to accomplish reliable performance. In the identification of the 'hidden competence' of the staff members this study attempts to offer a new perspective on the notion of 'competence'. This study aims to provide a multi-level analysis, the objective of which is to identify how a treatment trajectory is preserved in uncertain circumstances in ways that are frequently unrecognized or misunderstood without that analysis.

---

13   See, for the concept of 'exnovation', De Wilde 2000.

My analysis of the treatment trajectory shows how protocols (regulations) and numerical data (norms) have only a limited potential when it comes to facilitating a sound decision process in the NICU context. (For alternative discussion of the interaction between work practice and protocols and norms, see Chapter 3.) Numbers do not speak for themselves as they can also be the very cause of uncertainty. Likewise, medical progress and technological advances cause the boundaries of medical intervention to shift all the time, which underscores the significance of having protocols and guidelines. This idea is similar to that of 'drift' discussed by Wackers, in Chapter 4. However, this analysis also stresses the need for an overall awareness and professional attitude that neither regulations nor norms can entirely prevent vulnerabilities, near-misses or outright errors. A highly complex technological system can never become flawless, no matter how precise or elaborate its design may be. It is vulnerable by its very existence. Neither can the people who work within the system be labelled as its weakest link. They make it as much as they break it. This study emphasizes the necessity of space for adjustments based on what goes on in actual practice. More recognition of knowledge based on clinical experience seems vital. Although this form of knowledge cannot be formalized, it is no less accurate and wanted. If, however, medical practice is mainly governed by the fear of lawsuits, formalized forms of knowledge like protocols and quantitative standards will always play a dominant role. In these cases it is not only therapeutic reasons that predispose medical staff to collect numerical data. Legal conventions and the risk of a lawsuit occasion the collection and recording of numerical data and give rise to an avalanche of numbers. Numbers can serve as evidence in court to prove a physician made the most appropriate decisions. In this way other forms of knowledge, like the observations of clinicians, will have less space to develop, let alone be applied.

To ensure the continuation of the various processes that are an integral part of the treatment trajectory, more is needed – it turns out – than the often-heard plea for more advanced technology, stricter protocols and additional quantitative data. Having the opportunity to act circumstantially and make compromises is crucial for managing day-to-day operations in a complex technological system such as a NICU. More formal rules and regulations cannot rule out the need for these compromises. The dynamics of a complex technological system like a NICU eschew rigour, sternness and harshness. Instead, protocols have to be negotiable, facts have to be malleable (similar to the concept of plasticity discussed by Béguin, Owen and Wackers, in Chapter 1), and time has to be made. Only in this (sometimes) contradictory but pragmatic fashion can the medical staff construct its own point of departure upon which to base its daily activities and plan its patients' trajectories.

References

Anspach, R.R. (1993), *Deciding Who Lives: Faithful Choices in the Intensive-Care Nursery* (Berkeley, CA: University of California Press).

Atkinson, P. (1984), 'Training for Certainty', *Social Science and Medicine* 19.9: 949–56.
Berg, M. (1997a), *Rationalizing Medical Work: Decision-support Techniques and Medical Practices* (Cambridge, MA: MIT Press).
Berg, M. (1997b), 'Problems and Promises of the Protocol', *Social Science and Medicine* 44.8: 1081–88.
Berg, M. (1998), 'Order(s) and Disorder(s): Of Protocols and Medical Practices', in Berg, M. and Mol, A. (eds.) (1998), *Differences in Medicine: Unravelling Practices, Techniques, and Bodies* (Durham, NC: Duke University Press).
Bowker, G.C. and Star, S.L. (1999), *Sorting Things Out: Classification and its Consequences* (Cambridge, MA: MIT Press).
Brodwin, P. (ed.) (2000), *Biotechnology and Culture: Bodies, Anxieties, Ethics* (Bloomington, IN: Indiana University Press).
Brown, N. and Webster, A. (2004), *New Medical Technologies and Society: Reordering Life* (Cambridge: Polity Press/Blackwells).
De Wilde, R. (2000), 'Innovating Innovation: A Contribution to the Philosophy of the Future', *Paper for the POSTI Conference on Policy Agendas for Sustainable Technological Innovation, 1–3 December 2000* (London).
Franklin, S. and Lock, M. (eds.) (2003), *Remaking Life and Death: Toward Anthropology of the Biosciences* (Sante Fe: School of American Research Press).
Franklin, S. and Roberts, C. (2006), *Born and Made: An Ethnography of Preimplantation Genetic Diagnosis* (Princeton, NJ: Princeton University Press).
Frohock, F.M. (1986), *Special Care: Medical Decisions at the Beginning of Life* (Chicago: University of Chicago Press).
Garfinkel, H. (1967), *Studies in Ethnomethodology* (Englewood Cliffs, NJ: Prentice-Hall).
Guillemin, J.H. and Holmstrom, L.L. (1986), *Mixed Blessings: Intensive Care for Newborns* (Oxford: Oxford University Press).
Katz, J. (1984), *Why Doctors Don't Disclose Uncertainty*. The Hasting Centre Report (February): 35–44.
Latour, B. (1987), *Science in Action: How to Follow Scientists and Engineers through Society* (Cambridge, MA: Harvard University Press).
Law, J. (1994), *Organizing Modernity* (Oxford and Cambridge, MA: Blackwell).
Lock, M. and Gordon, D. (1988), *Biomedicine Examined* (Dordrecht: Kluwer Academic).
Lock, M., Young, A., Cambrosio, A. and Harwood, A. (eds.) (2000), *Living and Working with the New Medical Technologies: Intersections of Inquiry* (Cambridge: Cambridge University Press).
Lynch, M.E., Livingston, E. and Garfinkel, H. (1983), 'Temporal Order in Laboratory Work', in. Knorr-Cetina, K.D. and Mulkay, M. (eds.) (1983), *Science Observed* (Beverley Hills: Sage).

Mesman, J. (2002), *Ervaren Pioniers. Omgaan met Twijfel in de Intensive Care voor Pasgeborenen* (Amsterdam: Askant).

Mesman, J. (2005), 'Prognostic Differences and their Origins: A Topography of Experience and Expectation in a Neonatal Intensive Care Unit', *Qualitative Sociology* 28.1: 49–66.

Randell, R. (2003), 'Medicine and Aviation: A Review of the Comparison', *Methods of Information in Medicine* 42.4: 433–36.

Rapp, R. (2000), *Testing Women, Testing the Fetus: The Social Impact of Amniocentesis in America* (New York: Routledge).

Strauss, A.S. et al. (1985), *Social Organization of Medical Work* (Chicago: University of Chicago Press).

Timmermans, S. and Berg, M. (2003), *The Gold Standard: The Challenge of Evidence-Based Medicine and Standardization in Health Care* (Philadelphia: Temple University Press).

Tucker, A.L. and Edmondson, A. (2003), 'Why Hospitals Don't Learn From Failures: Organizational and Psychological Dynamics that Inhibit Systems Change', *California Management Review* 45.2: 55–72.

Wiener, C.L. (2000), *The Elusive Quest: Accountability in Hospitals* (New York: Aldin de Gruyter).

Wiskerke, J.S.C. and van der Ploeg, J.D. (eds.) (2004), *Seeds of Transition: Essays on Novelty Production, Niches and Regimes in Agriculture* (Uitgeverij: Van Gorcum).

Zussman, R. (1992), *Intensive Care: Medical Ethics and the Medical Profession* (Chicago: University of Chicago Press).

Chapter 6
How do Individual Operators Contribute to the Reliability of Collective Activity? A French Medical Emergency Centre

Jacques Marc and Janine Rogalski

Sociotechnical systems involving risk call for both reliable technical functionality and reliable human activity. Research was first focused on how to reduce – if possible to zero – human errors, within a common approach for technical and human reliability. Another orientation was to consider error as a normal component in human activity, which is linked to flexibility to tackle unexpected situations (the idea here is similar to what Messman (Chapter 5) calls space, and what Béguin (Chapter 7) calls 'plasticity'). This point was stressed in the case of aviation (see, for example, Maurino et al. 1995). In the case of aviation maintenance, Captain Maurino defended the approach proposed by cognitive psychology that 'error is a conservation mechanism afforded by human cognition to allow humans to flexibly operate under demanding conditions for prolonged periods without draining their mental batteries' (Maurino 1999, 4).

Amalberti developed a model of 'ecological safety' for analyzing human activity in dynamic and risky situations (Wioland and Amalberti 1999; Amalberti 2001a, 2001b, 2004) which integrates all the cognitive mechanisms through which operators ensure 'situation mastery' and perform 'sufficiently well'. A key component of this model is the process of 'cognitive compromise' for reaching this twofold goal with an appropriate allocation of cognitive resources. The model is linked with a shift of perspective in the issue of system reliability from error management to risk management (Amalberti 2004).

The hypothesis of cognitive compromise as a key process was used to analyze individual activity in risky dynamic environments (Doireau et al. 1997; Wioland and Amalberti 1999). This chapter questions how this approach may be pertinent for analyzing activity in collective work. We want to question how individual operators contribute to the reliability of collective activity through their error management. A second strand to our questioning is whether the ecological safety approach is applicable in this case of collective work and if the hypothesis of a cognitive compromise is supported. Hoc and Amalberti (2007) elaborate on the processes underlying the cognitive control on which the model is based. They specify two dimensions in these processes: the level of control (symbolic or sub-symbolic) and the origin of the data used for control (internal–anticipative, external–reactive) in the case of individuals. In the case of collective work, when individuals act as supervisors of others' activity, we consider that the cognitive control is symbolic and external–reactive.

In a study of a medical emergency regulation centre (a French telephone call/response centre, 'Service d'Aide Médicalisée d'Urgence', SAMU), our field observations explored the relevance of the model, and particularly tested the fact that not all errors are identified and recovered, while the safety level remains high (Marc and Amalberti 2002). They enabled identification of a set of cases of error management which was used to design a simulation, and to analyze how operators managed errors when placed in a position of observers of the activity of their own workplace, that is, at the SAMU. The simulation enabled confrontation of the same situation with actors playing different roles (physicians and call-takers) in the SAMU. It also enabled the observers to verbalize what they observed in the simulated situation, and their assessment of the mastery of the situation and level of risk. Such interrogations were impossible in the real setting for safety reasons. In these studies, we analyzed the role played by observers in facing collective work with regards to operational memory and shared situation awareness, the status of possible departures from prescribed procedures (are they or should they really be considered as 'errors'?), and attribution of agency (individuals, roles, a group as an entity) in events related to safety. For the relevance of the model of ecological safety to be demonstrated in the case of collective action, some development needs to be made to the mechanism of what has been called 'cognitive compromise'.

Risk, Safety and Error

A risk is an event, specified by two attributes: its possibility of occurrence and the negative impact of its outcomes. Its occurrence is neither certain nor impossible, and cannot always be assessed in probabilistic terms; its outcomes may involve quantitative issues such as loss of production in volume or money, and/or repairable material damage, or qualitative outcomes, including injured or dead people and/or irreversible damage. Safety is often assessed through the consideration of occurrences of risks; this may lead to paradoxes as underlined by Amalberti (2001a).

Errors can be considered from the point of view of the task (the tasks requirements) or from the point of view of the operator's activity (their goals). In highly proceduralized situations in risky work, any departure from the prescribed procedures is considered an error; it was the case in Wioland, Doireau and Amalberti's studies in aviation, and in the empirical study analyzed in this chapter. Such a definition expands the 'level' of error: skill-, rule- or knowledge error, or its 'nature': real error or violation of rules, as analyzed by Reason (1990). As soon as expert operators are involved and situations are not exceptional ones, knowledge-based as well as rule-based errors are quite infrequent, and the distinction between error and violation of rule does not matter from the point of view of departure from procedures.

Human errors may trigger a sequence of events leading to the realization of some feared risk, and hence they threaten safety. As for technical faults, not all errors result in an accident or incident. Particularly in dynamic environments, changes in context may 'cancel' what was previously an error, or lead to minimal consequences.

Elsewhere, errors may be considered as cognitive resources: occurrences of an error indicate the operators are reaching their limits. In other words, errors may be determined by their own activity (acting too fast, tackling too many tasks in parallel, not using support tools, among others), or by their state (tiredness, insufficient sleep) or eventually competence. Errors may also derive from the situation (too much information, excessive work load, increased complexity, other actors interfering).

Error management concerns all cognitive activities and material actions whose object is error itself: it includes prevention, detection, identification, evaluation, supervision and recovery. It may also concern the use of errors for assessing operators' activity limits. The classical approach of reliability through a focus on error management is not the only possible approach. For instance, Hollnagel (2005) proposed an approach of 'performance variability management', while an approach in terms of 'risk management' may be focusing on the resilience of a system when facing surprises, leading to crossing the boundaries of safe operations.

Risk management concerns all activities oriented toward risk limitation, either from the point of view of its occurrence (prevention) or of its consequences (recovery actions, anticipation and organization of required resources). Situation assessment – both with regards to the dynamic event and to the resources available with which to face it – was shown to be a critical competence in managing a dynamic risk (Rogalski 1999). Moreover, professionals (fire-fighters' superior officers, in this case) were more often centred on risk consequences than were naive subjects (such as mathematics and psychology students).

Ecological Safety and Cognitive Compromise

Risks are manifold, in that they are related to various goals. Some of these risks are external risks: for instance, in the case of civil piloting, they concern problems in aircraft safety that may lead to human injuries or death, and/or material damage. In the case of medical emergency regulation, external risks mainly concern the patients' lives. External risks also have an organizational dimension: pilots going beyond a formal rule in the case of an abnormal situation might lose their pilot permits. Internal risks deal with the actors' cognitive activity itself. Compromises intervene both within the set of external risks and the hierarchy between them, and between external and internal risks.

These points will be developed below, in the case of SAMU 75. Its telephone regulation centre has to cope with various patients' calls. Some are vital emergencies: any delay in sending rescue personnel may be fatal to the patient. Some other calls do not constitute a real emergency, despite the patient's or relatives' anxiety: they are false alarms with regards to emergency decisions; they need other types of help. The main purpose of the SAMU 75 telephone regulation centre is to ensure that threatened lives will be preserved through rapid intervention. As emphasized by M. Mort (personal communication), their activity involves 'triaging'; that is, making judgements about the severity and urgency of the case. The triage nurse in an accident and emergency (A&E) department does the same. The actors in the telephone regulation centre have to do 'telephone triage', which requires a particular skill, requiring particular training and involving much non-explicit, or tacit, knowledge.

External risks involve both the current task of responding to a specific call for emergency help and the long-term goal of ensuring the availability of resources (ambulances and mobile intervention units that enable immediate resuscitation and patients' preparation for transportation). Assessing the first external risk is a difficult point because the actors in the telephone centre do not have any direct access to the patient's state, and have to identify the severity of the situation and emergency from callers who may be without specific medical knowledge and/or are often under stress (the patient's relatives, for instance).

Internal risks concern the actor's activity itself; in collective work, internal risks are also related to cooperation activities, and to the group's activity, where the group is considered as an entity. The main idea is therefore to consider a group such as an emergency call centre as if it were a real (collective) subject, without considering who is involved and what is the group or organization, but only what is the task that is prescribed to the group and how the group performs it. As it is not a real (human) operator, but rather a concept concerning the overall collective work, we called the group entity, a 'Virtual Operator' (Rogalski 1991). Concepts and methods developed for analyzing individual actions can be applied to the Virtual Operator. However, this raises the issue of whether all cognitive processes that are postulated for individual operators can be transposed in the case of such a Virtual Operator: it is particularly the case for the mechanism of cognitive compromise postulated in the model of ecological safety.

The cognitive compromise is a central mechanism postulated in Amalberti's model of ecological safety. Schematically, cognitive compromise can be described as follows: in managing potentially conflicting internal and external goals (and associated risks), operators try to ensure an 'acceptable' operational performance (external risk as a counterpart) while staying in control from the point of view of their representation of the situation (internal risk as a counterpart). Two consequences follow. On the one hand, operators tend not to be overwhelmed with cognitive activities oriented toward assessing the situation in order to be aware of its main features, and they aim at acting 'sufficiently' well, while accepting 'impasses' (in some cases, people may 'skip' parts of their tasks as when someone 'finesses' in the game of bridge). On the other hand, they may commit errors, as long as their consequences do not threaten the state of the immediate future situation, and as long as the errors are not irreversible. From the point of view of risks, external risk is present in possible negative consequences of operating at a minimal acceptable level, while internal risk comprises having a limited mental model of the occurring situation.

The model of ecological safety proposed that operators make compromises between these two risks (external and internal), try to establish a representation of the situation 'sufficient' to be useful for planning and controlling action, and aim at operating in a way which is not the best but which allows them to keep external risk within acceptable limits[1] (see, for example, Wioland and Amalberti 1999). The model was developed in contrast to approaches of safety and risk management, which are mainly centred on error avoidance and directed towards the system's overall performance. 'Ecological safety involves the set of mechanisms an operator makes use of, for ensuring situation mastery and enabling a sufficient performance. It may be a matter of error detection and recovery, of strategy for controlling risk taking, etc.' (Amalberti 2004, 80, our translation). The concept of cognitive compromise is one of these mechanisms.

The notion of 'sufficiency' underlying good enough performance 'expresses a property of the cognitive system which enables operators to complete their work without reaching the maximum performance. It is involved in all the domains of control: planning, decision-making, and execution' (Amalberti 2004, 78). Elsewhere, Amalberti underlines the role of situation mastery as being:

> at the core of dynamic situation management for it allows us to understand how the dynamics of cognitive processes are adapted to the dynamics of the controlled process. From this concept follows the notion of 'understanding sufficiency' for acting while ensuring that the cognitive resources utilised and process control remain within acceptable limits. (Amalberti 2004, 107, our translation)

This notion does not only express a limit in the representation process. The importance of a 'minimal' situation assessment for remaining in control was observed in a study of how pilots face 'automation surprises' (unexpected behaviour by automated systems), in spite of the fact that there exists a strictly prescribed procedure – that is, 'go back to basics' and to immediately disconnect the automated systems concerned (Plat 2003).

The notion of 'consistency' (Marc 2002) was introduced in order to be precise regarding this quality of knowledge about the situation: consistency means that the subject's representation of the occurring situation is not only 'sufficient' (in its extension) but also coherent enough (in its organization) for the purposes of their present action (operators may trust the operational validity of their models of the

---

[1] A similar notion was introduced by Simon (1982) in the domain of decision-making in organizations, under the term 'limited rationality'.

situation). The quality of the compromise between external and internal risks, as postulated by the model of ecological safety, depends on situation awareness.

Error and Safety Management in Collective Activity

The ecological safety model and the notion of cognitive compromise were developed and studied in the frame of individual activity. In numerous situations, operators are interacting in collective work, with more or less interdependent tasks. Does the model of ecological safety remain valid and how does it need to be accommodated to tackle the issue of collective work? Two points of view may be taken: (1) one considers how the various individuals interacting in collective work contribute to safety management; or (2) one considers the system of actors involved in collective work as an entity. Von Cranach et al. (1986) used the term 'a group as a self-active system'; Rogalski (1991) introduced the notion of 'virtual operator' in order to transpose concepts and methods used for analysis of individual activity to collective activity. To do so, we cannot simply apply the concepts as such; rather, they need to undergo some transformation because an individual is a physical unit, whereas a Virtual Operator is not.

From the first point of view – that one considers how the various individuals interacting in collective work contribute to safety management – an example of analysis of errors in collective activity can be found in Jentsch et al. (1999). A detailed analysis of more than 300 incident reports in civil aviation indicates: (1) that the person who is flying commits more errors that lead to an incident, and that the non-flying pilot is less likely to lose situation awareness; and (2) that captains (who have the legal responsibility for the flights) lose situation awareness more often and make more tactical errors when they are at the controls than when they are not. Team role assignment thus appears as a twofold variable in errors leading to incidents. First, with regards to process control, front-line actors are more prone to lose situation awareness than persons who are in more of a position of being observers. Second, persons in charge of team responsibility make fewer errors when they are not directly engaged in controlling the process.

The second point of view regarding the validity of the model of ecological safety – that one consider the system of actors involved in collective work as an entity – is taken up in a study aiming at identifying cues of expertise in collective work in public safety, in a Command Post of firefighter officers in which Antolin-Glenn analyzed a critical event of problem-solving (Antolin-Glenn and Rogalski 2002). The issue was quite efficient, but there appeared to be many ambiguities, instances of non-precise information, and errors (slips) in identification of information or computing, although there was always somebody in the team to recover them. In fact, many interactions were oriented towards shared situation assessment, enabling recovery of errors through crosschecking – actors mutually checked each other's information assessment and operations (Rogalski 1996).

Collective Work in the SAMU 75

Medical emergency regulation centres in France are organized on a common basis. Physicians are responsible for interaction with callers in order to diagnose how urgent the medical problem is that motivates the distress call. The call-takers, known as Permanenciers Auxiliaires de la Régulation Médicale (PARM), are in charge of the first interaction and can play a filtering role (to a degree) when it is clear that there is not an urgent medical problem, for instance, calls from drunken people or hoaxers. Call-takers also give 'missions' (instructions, orders) to the medical ambulances known as Service Mobile d'Urgence et de Réanimation (SMUR), while a radio operator is responsible for a specific communication network (alternatively, this task may be devoted to a call-taker). More generally, call-takers are in charge of external interactions, with the exception of medical matters. The basic composition in SAMU 75 is that there are two call-takers (one of them is in charge of the radio), and one physician. Depending on the workload (expected emergency calls), there may be more call-takers and physicians in the regulation room (up to six call-takers and three physicians, one physician specializing in child care). It is up to the physicians to ensure the quality of teamwork: in this respect, the physicians play a role similar to those of captains in civil crews, to the Chief in a Command-Post (Antolin-Glenn and Rogalski 2002), or to the supervisor in a control room.

Activity is organized around cases (in French: '*affaires*') that constitute units of action: a call triggers case management. Each case is linked to a record in a file that is opened by a call-taker, which will also be used by a physician and constitute both a tool for information sharing and a common object of action (filling it as required is an administrative and legal obligation). In normal situations, case management involves a number of micro-procedures concerning the patient, the file, and internal communication. The first actions to be undertaken are as follows. The call-taker takes the call, opens a file, locates the call, performs an initial sorting and emergency assessment, 'informs' the file, informs the physician, and closes the file while transferring the call with the file number to the physician. The physician takes the call, opens the file, assesses the medical emergency and, if possible, the diagnosis, fills in the medical record fields, chooses a means of action, writes down this information, informs the PARM about their decision, closes the file, and if necessary relays the call to the PARM. Action continues until the patient has been taken in charge, and the case file is closed. Any departure from the requested organization of such micro-procedures is considered an 'error': not all of these departures or 'errors' may result in threatening the patient's safety, but all may have administrative, financial or legal consequences. These activities comprise the basis of the data that we analyzed.

External and internal risks in a system of collective action

The main risks involved in the regulation centre are presented in our case study. Their generality is discussed from the point of view of the (possible) transfer of both issues and results to other situations of similar types (particularly non-profit rescue, medical or social services).

As stressed by Rasmussen (1997), risk management involves the whole work system. Analyzing collective activity of the front-line actors in the regulation centre enables us to tackle only part of risk management. However, taking into account the whole set of objectives that SAMU is aiming at enables us to establish links between organizational levels. In effect, the first objective is an operational one: responding to the callers' medical emergencies. Another objective is tracing operational activity for legal reasons, because of the responsibilities the SAMU has to assume. A third objective is tracing operational activity for administrative and financial purposes: justifying that the expected activity has been performed (a posteriori justification) and proving that there are further resources to meet demands (a priori justification).

External risks concern, on the one hand, the 'objects of current action' (the patients calling for medical help, which is related to the first and second goals), and, on the other hand, the long-term conditions of action (which are related to the third goal). Internal risks are linked to goals oriented towards the actors' activity itself: they concern an individual's activity, interferences with other actors (provoking as well as preventing or recovering errors), and activity of the group as a whole. They involve the risk of losing control of the situation in elaborating an insufficient occurring representation, or in giving too much time to a specific case to the detriment of other and perhaps more crucial representations.

One point has to be stressed: there is no one-way link between actions performed by SAMU and types of risks involved. For instance, filling in the 'localization' field in a case record is not a risk from an operational point of view if the case appears not to be an emergency (a drunk person, for instance) but the risk is an administrative and financial one; the same omission may have serious consequences if communication is cut for any reason while the case proves to be an emergency. The type of external risk depends on the case, whose quality is not known when such an omission is eventually made. However, not opening a record in the case of a non-emergency call limits the risk of wasted time and attention, when numerous calls are expected (due to the date or the particular moment).

In collective action, individuals may detect errors committed by other actors, as shown in a study by Doireau et al. (1997) (observers – particularly if they are experts – are able to detect many errors). In this sense, error detection may decrease external and internal risk (losing situation awareness, for instance). However, if error recovery requires that its producer becomes aware of their error, and hence if a specific communication has to be performed, it can both produce a negative interference in the producer's flow of action and use temporal and attention resources, thus increasing the internal risk of overload. Moreover, interacting with others about a committed error may also increase the risk of distracting the attention of the one who identified the error.

As a consequence of the model of ecological safety applied to collective work, such errors – committed by an individual in a group – may be detected by another actor more often than when concerning their own error. More precisely, what could be expected is that an individual more easily detects others' errors, but among detected errors, more frequently recovers their own. Moreover, actions (mainly communications) preventing future errors in collective work should occur frequently, with regards to detection and recovery. In general, such anticipation aiming at preventing errors remains non-observable for individual activity.

Individual Error Management in Collective Activity

How does an individual intervene for safety management in collective action? A study detailed in Marc and Amalberti (2002) focused on how an individual actor manages risk and safety both from the point of view of their own activity and from the point of view of taking part in a collective activity. Over two months, the activity of an experienced call-taker in the regulation centre was systematically observed. The cases on which information was complete were the subject of an error analysis. Every departure of the team from the expected micro-procedures was noted as soon as it could have an impact on the call-taker's activity. Each such departure was categorized both concerning the nature of error (routine error, knowledge errors, or violations; see Reason 1990) and concerning its actor (the call-taker herself or himself, whether alone or interacting with somebody else; another actor; or the group itself). The call-taker's actions for preventing errors occurring in the group were also noted. Thus, the call-taker's activity was analyzed with regards to detection, recovery and/or prevention of error.

Globally, for the events under scrutiny, the analyst identified 61 per cent as errors (of commission or omission) and 39 per cent as preventing actions. Among the errors, only 27 per cent were detected and recovered by the call-taker (in fact, these occurred in cases where the call-taker was involved as the producer or as a team member), and 17 per cent were detected without recovery operations (these were mainly individual errors). Most of the errors detected without a specific recovery action were 'erased' through the normal collective activity, although nobody detected them. Routine errors, knowledge errors and violations were almost equally represented. Knowledge errors that were detected were mainly committed by other team members. These data meet results already observed for individual error detection: for instance, experts learn by experience that slips (routine errors) are not detected in a standard way (through online identification) but are more often detected by checks (Allwood and Montgomery 1982; Rizzo et al. 1987); expert observers are also able to identify knowledge errors (Doireau et al. 1997), while these are difficult for the producer to detect themselves. Finally, most of the detected errors concern short-term consequences (such as correcting erroneous information during management of a case), while prevention actions typically concerned team activity (such as stressing the risk of lack of resources for future cases, or the risk of a possible excessive workload with several cases to be managed simultaneously).

The fact that many prevention actions were oriented towards other actors or the group itself reveals the role individuals play in collective safety. Moreover, in spite of the fact that many errors (departures from what is prescribed) were not recovered, the overall performance of the regulation centre was quite accurate, in that none of the observed errors were detrimental to medical efficiency. From our point of view, several elements are involved. First, not all errors concerned the medical facet of the SAMU activity: many of them were related to administration (filling in the records in the file). Second, it is reasonable to hypothesize that error management in collective work is focused on a relatively high level of errors in terms of their potential consequences, in order to invest only a few resources in supervising, detecting and recovering errors, unless they threaten (medical) efficiency. Third, SAMU's actors were quite involved in preventing overload and/or a lack of resources. Lastly, all of the activity happens as if ensuring shared situation awareness was a crucial (and eventually sufficient) condition for avoiding severe errors.

Assessing individual and collective errors

The case study indicated that when individuals are involved in a collective action, they manage errors in a way that is compatible with the existence of a cognitive compromise as a twofold mechanism acting at the individual level and at the level of the group as an entity (a Virtual Operator). Not all errors are detected, some are detected but without correcting actions; action is also oriented towards the collective mastery of the situation (both through supervising and preventing errors).

An experimental design was used to: (1) test the existence of shared mental models of the occurring situation – such shared mental models being at the basis of the Virtual Operator activity; (2) identify if a hierarchy of goals underlies the mechanism of cognitive compromise in the case of collective action, such a hierarchy being identified through differences in error management depending on their object; and (3) assess to what extent allocation of errors (to individuals or the group as a whole) was reflecting the reality of error production (an issue meaningful for collective action), and was linked to evaluation of the mastery of the situation (detected operators' errors being cues of their acting 'at the limits').

In this experimental situation where SAMU actors were put in the position of observers of a (simulated) phase in SAMU activity, the method was transposed from the initial study by Wioland and Doireau (1995) that was centred on individual activity. Some adapted changes concern particularly the concept of the regulation centre as an entity, beyond the consideration of its individual actors, and introducing issues about cooperation within it. Due to the variability of the observed situations in the SAMU activity, and due to confidentiality issues preventing activity records, an experimental situation – a simulation – was designed.

In collective work, individuals may be considered as occupying (at least) two positions: they are actors in individual and collective action and they are observers of the activity of other actors and of the group as a whole. (In this last position they can neither produce errors nor recover others' errors, but they may detect, identify and anticipate errors.) In the simulation there was a 'decoupling' (Samurçay and Rogalski 1998) of these two positions, as the professionals were not allowed to act, neither on the 'objects' of the situation, nor on the SAMU's activity. This has to be considered a first step for assessing the relevance of the model of ecological safety in the case of collective work, as the model would concern operators in their various roles.

The simulation was based on the previous study, with errors in SAMU activity effectively observed. The two types of professionals – 13 call-takers and 13 physicians, almost all of the personnel of SAMU – were confronted with the same simulated situation, involving two call takers and a physician. They were asked to observe the activity of operators in their centre (played by real professionals in the SAMU), and to comment – during (pre-defined) breaks – on what they observed. Finally, they were asked how they would assess the mastery of the situation and the level of explicit and latent risk in the SAMU. The position of observation was 'above the shoulder' of a call-taker (PARM).

Eight cases were introduced in the simulation. They involved 49 episodes where an activity developed in the SAMU concerning safety: individual errors and violations of the various types observed in the previous study, individual or collective activity, organization that might lead to errors (interferences, parallel task performing, lack of attention to others' needs, and others), and also occurrences of 'defences' (positive defences were preventing actions and error recovery). Twice as many errors as defences were introduced in the scenario. Half of the safety episodes concerned one PARM (PARM 1), while the remainder concerned equally the PARM and the physician (PARM 2).

Operational Memory and Shared Mental Models

Comments about what they had observed when watching the simulated situation show that all professionals, whether physicians or call-takers, memorized quite well both the 'patients' stories' (cases), and the activity developed in the SAMU: 'Operational memory' (Bisseret 1970; 1995) was quite high in the two groups. Comments were mainly related to direct safety issues: for instance, they did not mention a case that did not involve the patient's health risk – although the physician was performing two tasks simultaneously, with potential consequences for his future activity in the scenario. Operational memory was also attested to in the fact that comments on some initial cases were continued all through the simulation session, while relationships between episodes were considered almost as often as individual episodes were referred to in isolation (as specific safety facts). The high quality of operational memory allows us to analyze the existence (or absence) of detection of the errors introduced in the simulation script, and to analyze comments on the recovery or protection actions, as being related to error management – and not to insufficient memorization of what happened in the scenario.

Similarities in the results from the two groups of SAMU actors also indicated that mental models about the current situation (in the SAMU) were widely shared. The notion of 'shared mental models' was used by Rouse et al. (1992) in the case of aviation crews. It is related to the notions of 'shared situation assessment' and 'shared situation awareness' developed by Endsley (1995) for individual activity. In fact, shared mental models also presuppose common work process knowledge (schematically: about the objects of action and about the actors' activity), which was considered as a condition for efficient collective work under the term of 'operational common system of references'.

The fact that shared situation assessment was quite high also confirms the interpretation we proposed for the greater place of prevention actions with regards to recovery actions which was observed in the previous field study. However, some differences were also observed in memory of episodes, in relationship with the roles within the SAMU: physicians more often referred to episodes involving the activity of the observed physician and management of high-risk cases, while call-takers were aware of errors involving file management (one of their main duties) more often than were physicians.

Managing Safety

There were several types of comments about safety: anticipating (possible) errors in (simulated) SAMU activity; detecting and identifying errors; identifying positive defences or noticing 'negative defences', that is, the absence of (expected and possible) error recovery; proposing SAMU actions in a counterfactual way; referring to the trajectory of errors during the scenario (consequences of errors, or 'natural error death' as when a case appears not to be an emergency, or when other actors manage it). In fact, most of the comments about safety were about detecting errors (almost 90 per cent).

Several reasons may explain this result: detecting error is a necessary cognitive step in error management, but it does not imply the presence of further steps; the absence of proposal of 'counterfactual' actions could have resulted from a general orientation toward 'prospective' actions for error recovery and not 'retrospective' criticism as involved in proposing 'counterfactual actions'. There was little anticipation among the comments, and these were mainly proposed by physicians: this could be related to their tactical role (similar to that of captains in pilot crews), that is, the necessity to consider the long-term team activity'. Referring to the trajectory of errors could also be related to tactical considerations.

Not all departures from the prescribed procedures were managed as errors. Globally, the SAMU members first focused on medical regulation errors (involving patients' safety), then on positive defences. Errors concerning management of files were reported by fewer persons. Comments concerning interferences, where the object of error (by commission or omission) was cooperation itself, were few, as were occurrences of inappropriate situation assessments related to patients' state (which constitute latent errors more than patent ones). This result indicates a hierarchy of goals and associated risks that might be inferred from the differences in safety (and error) management depending on the objects on which safety is concerned: the main goal was the operational one, related to patients' safety (regulation itself); administrative-oriented goals (such as file management) were assessed quite below the first and main goal. These external risks could all be assessed through what was done on the operational objects of action: cases involved in emergency calls and records in the file. It is not so easy to comment on interference errors from this point of view, as to do so clearly requires speaking about activity itself (and not its operational objects). Could this be why interferences were more poorly considered, due to a certain lack of reflexive habits with regards to operational interferences? This hypothesis is congruent with the absence of a proposal of 'counterfactual' actions: a lack of reflexive habits about the team's activity could lead to interpret such a proposal as criticism and not as a way for improving collective efficacy as regarding risk management.

Assessing Situation Mastery

Most detected errors were allocated to specific actors (only 13 per cent were referred only to their object, and 3 per cent were referred to people outside the regulation centre). Within the regulation centre, allocation concerned the physician, an interaction between the physician and call-takers, an individual call-taker or an interaction between the two call-takers, or the regulation call centre as an entity (our Virtual Operator).

There were very few differences between physicians and call-takers with regards to the allocation of errors; but more call-takers than physicians considered the physician to be involved in error production through the interactions with call-takers, while more physicians considered the regulation centre as being involved in errors production.

In fact, error allocation did not follow the distribution of errors as introduced in the simulation scenario, in which a specific actor initiated all errors. Interactions between actors (between the physician and a call-taker or between the call-takers) and the activity of the regulation centre as an entity comprised about one-third of all error allocations (error allocations were almost equally distributed). Moreover, while PARM 1 was the main individual error-producer in the scenario (responsible for producing about half of the errors), PARM 1 was not allotted more than 13 per cent of all errors (the same percentage as the SAMU group as an entity); on the other hand, while the physician was involved in only one-quarter of the errors, the observers allotted more than 45 per cent of direct and interaction errors to the physician.

Physicians and call-takers also agreed in assessing the situation mastery shown by the individual actors and the group itself: they assessed the lowest mastery to the physician (less than 2.5 points on a 7-point scale), followed by the group itself (3.3), with call-takers' mastery assessed as average (slightly above 3.5), without significant differences between the two call-takers.

These distortions indicate that assessing collective safety management has specific features, and cannot be identified as a collection of assessments of individual activities in the group. First, the group was considered as a Virtual Operator, both from the point of view of error allocation and situation mastery. Second, the person in charge of the collective work quality – the physician in the regulation centre case – was considered to be the key individual in safety management.

Conclusions

Globally, data show that individuals contribute to the reliability of collective activity through process-sharing characteristics of the properties involved through individual error management. They also stressed specific points linked to the collective dimension, related to the articulation of individual activities in the group action, to the existence of the group as an entity, and to the specific role played by the physician.

There appeared strong convergences between the results of observations in the SAMU regulation centre and the data issued from the simulation. In the first case, in their real professional setting, team members were both observers of the others' activity and actors, in performing their own immediate tasks and in interacting to help achieve the overall goals of the SAMU. Errors were committed, without always being detected by their producers; while detected by another actor acting in an observer position, errors were not necessarily recovered through interaction. It was also observed that even if detected errors were not recovered, they were supervised until the case's final resolution. These points support the pertinence of the model of ecological safety, involving cognitive compromise as a key mechanism, in the case of individuals acting in collective work (both as actors and as observers). Moreover, there were various actions aiming at preventing 'conditions for errors' oriented toward others' activity and to the regulation centre as a whole.

Actually, individuals played several roles with regards to collective work safety: as actors, they were producing and recovering errors with regards to their own current task; as observers, they acted as supervisors of collective activity, checking if detected errors remain harmless in a case life, and preventing conditions for errors to occur, through interventions towards the group members to urge them to pay attention to internal as well as to external resources. There were also many interactions oriented towards sharing situation awareness: this could be interpreted as the importance for the group to elaborate on a representation to ensure consistent understanding within the group so that sufficient control of the situation is maintained, and such that cognitive resources are also engaged at a reasonable level. This could be interpreted as the existence of a mechanism peculiar to collective work, different from but coherent with the individual mechanism of cognitive compromise.

In the simulation, interaction with the observers led to converging conclusions: memorization of cases and episodes of the scenario was similar to memorization observed in real settings (initial problems being kept in mind until their final resolution), and attested that individuals assessed the situation with a large community of representation. Errors and safety management were also in line with previous observations: not all errors were detected; a hierarchy of goals could be inferred from the differences in detection in that operational errors (with possible immediate effects on patients' safety) were the best detected, while more long-term errors, related to administrative goals, were less often stressed. Collective operational orientation could be inferred from the fact that negative interferences in the regulation centre were not frequently referred to (as long as they were not directly involving external risks). Moreover, defences against errors were also stressed by most of the subjects: that is, in the same line of safety management as the prevention actions observed in the real setting.

Even when strongly criticizing the quality of collective work observed in the simulation scenario – particularly from the cooperation point of view – SAMU members generally agreed about the overall final performance, and the ability to always recover from a risky situation, thanks to the level of medical resources available in the observed SAMU. 'The group managed. The individuals managed'; 'Each case received an answer. Perhaps not a good one; but there was no severe risk'. If some comments underlined the likelihood of a deteriorating situation, due to more critical cases or an acute vital emergency, others stressed, 'even in a multiplication of risks, it is always possible to recover'. These points can be interpreted in the light of the model of ecological safety and applied to collective work. In given situations, the SAMU system was reliable, in that it remained within the boundaries of safe operation. In this way, participants assessed that the system was resilient, that is, it would be able to act positively if the boundaries of safe operations needed to be crossed, thus activating organizational goals and actions to limit consequences and to return to a safe state.

In line with the ecological safety model, safety management in collective work appeared not to be error avoidance or error recovery. Safety management was, on the one hand, being sensitive to differential effects of errors: the main objective being to avoid major consequences, so short-term operational goals had first priority and administrative error management with no direct consequence was ranked lower. On the other hand, conditions that threatened efficiency were prevented, particularly through managing cognitive resources and thus limiting later workload.

Specific processes related to the 'collective fact' were also observed, leading to the addition of group-oriented mechanisms beyond the individual cognitive compromise. They concern the articulation between individual activities in group action, the existence of the group as an entity, and the specific role played by the physician, in charge of the global work quality.

In the real setting, detecting others' errors without intervening might be considered to be a mechanism similar to the cognitive compromise between external risks (consequences of errors) and internal risks (losing the cognitive control of the situation; being overloaded), applied to the collective situation, at a symbolic level. Recovering an error committed by other actors often requires capturing their attention (or interfering with their actions). As a consequence, a part of the cognitive resources of team members was devoted to management such an interference, to the detriment of their managing the case at hand.

Comments expressed in the simulation add another mechanism: that 'in a stressful situation, the trend is to manage cases as fast as possible' (which is similar to the individual model of compromises between external risks and internal risks – here, the risk of being overwhelmed by the number of cases), and 'to get rid of them through immediate passing on, without taking care of the colleague's emergencies'. Individual safety management may thus be conflicting with overall group safety management. Managing the SAMU cases calls for both individual and inter-individual parallel activities.

Another specific point concerns internal risks related to cooperation: cooperation may appear as a goal of its own, which, under temporal pressure and high workload, may conflict with short-term individual operational goals. Internal risks related to cooperation were noticed by several observers: 'it leads towards a tiring atmosphere', 'they happened to solve it, but they must have been exhausted by the end', 'this functioning will necessarily lead to break-ups'. It looks as if individual compromises of safety management prevailing over compromises involving cooperation, at least when there is no one taking the last compromise decision at an appropriate place (as was the case in the simulation). Similarly, in aviation, cooperation quality appeared to be related to operational mastery in studies on collective cockpit resource management (Plat 2003; Rogalski 1996).

Similar findings were noted in the case of crews confronting automation surprises (Plat 2003; Plat et al. 1998). It was shown that explicit cooperation is a process subordinated to situation mastery. Conversely, without a sufficient and consistent model of the situation, the primacy of actions to control the dynamic environment occurred at the detriment of explicit cooperation, which should contribute to build an appropriate situation assessment. The cognitive compromises that regulate individual activities (following Amalberti's model) became more complex, as they needed to involve cooperation in the case of collective work.

Other specific results concern safety management at the group level. First, the role of the physician as the actor tactically responsible for the collective work was emphasized through an overestimation of his implication in errors production. Second, the regulation centre was treated as an entity by allotting it a significant contribution in error involvement, to the detriment of individual operators' (mainly call-takers') effective contribution.

The physician's role was described (from the point of view of the observers) as that of a 'conductor of an orchestra', a role he was not performing in the observed simulation. The articulation between call-takers' and the physician's activity was also stressed, linked to two types of risks: the risk of insufficient availability of the physician and the risk that a call-taker orients their assessment of a patient's state when giving their own interpretation. Related operational risks concerned sending inappropriate resources, with consequences for the patient's health or life. Being the person in charge of the global performance and quality of the collective activity, the physician is then considered as having to regulate his own interactions with the call-takers, possible contradictions between individual goals and activities, and to ensure smooth dynamics within the group.

These two points respectively refer to the two perspectives we underlined for analyzing collective work: individual functions and articulation of tasks and activities in the group, and consideration of the group as a Virtual Operator. Moreover, the emphasis operators put on the simulation on the physician's activity from the point of view of the collective work leads to the proposal that the physician – as the person in charge of the global collective work – could be considered as personifying – in the real world – the group as a Virtual Operator – represented at the conceptual level.

The last issue concerns the ecological validity of the simulation and the domain of validity of the previous analyses. Two types of cues testify for the simulated situation being representative of real situations, considered from the point of view of an observer. First, we have already underlined how well the results from observations in the real setting converge with those coming from the simulation. Second, the subjects, physicians and call-takers, acting as observers of the simulated scenario, all stressed how realistic it was: 'It is quite a classical situation in SAMU 75 regulation', 'It's what we may encounter everyday', 'It's a situation I know, you did not invent anything'. They all also agree about the reasons why this was the case, the main one being insufficient SAMU personnel, and under-developed training to unify individuals into teams. Moreover, some of them were trying to act (during the simulation), searching for commands to correct files when identifying erroneous information about a case.

The situation upon which we based our analyses about individual and collective safety management can be considered to be a generic example of a variety of collective situations, where actors collectively interact to manage dynamic environments, under time pressure and with risky external consequences of their actions. Both the issues tackled concerning safety management, the methods used for the study, and the main results, could be applied to a class of collective situations, in which cooperation is organized both horizontally (task allocation to several actors) and vertically (a particular actor having the overall responsibility – like a conductor of an orchestra). We proposed to add collective-oriented mechanisms to the cognitive compromise postulated in Amalberti's model of ecological safety, linked on the one hand, to the group functioning as an entity – 'personified' by its responsible actor: the physician; and on the other hand, to an equilibrium to be reached between individuals' cognitive compromises. Such considerations are quite valid for the set of dynamic risky situations managed with a similar operational device, where actors are in the twofold position of potential observers and actors.

References

Allwood, C.M. and Montgomery, H. (1982), 'Detection of Errors in Statistical Problem Solving', *Scandinavian Journal of Psychology* 23: 131–39.

Amalberti, R. (2001a), 'The Paradoxes of Almost Totally Safe Transportation Systems', *Safety Science* 37: 109–126.

Amalberti, R. (2001b), 'La maîtrise des situations dynamiques', in Cellier, J.-M. and Hoc, J.-M. (eds.) (2001), *La Gestion d'Environnements Dynamiques. Psychologie Française* 46(2).

Amalberti, R. (2004), 'De la gestion des erreurs à la gestion des risques', in Falzon, P. (ed.), *Ergonomie* (Paris: PUF).

Antolin-Glenn, P. and Rogalski, J. (2002), 'Expertise in Distributed Cooperation: A Command Post in Operational Management', in Bagnara, S. et al. (eds.) (2002), *ECCE'11: Cognition, Culture and Design*. Catania: Instituto di Scienze et Technologie della Cognizione (CNR).

Bisseret, A. (1970), 'Mémoire opérationnelle et structure du travail', *Bulletin de Psychologie* 24(5-6): 280–94.

Bisseret, A. (1995), *Représentation et décision experte: Psychologie cognitive de la décision chez les aiguilleurs du ciel* (Toulouse: Octarès Editions).

Doireau, P., Wioland, L. and Amalberti, R. (1997), 'La détection des erreurs par des opérateurs extérieurs à l'action: Le cas du pilotage d'avion', *Le Travail Humain* 60(2): 131–53.

Endsley, M.R. (1995), 'Toward a Theory of Situation Awareness in Dynamic Systems', *Human Factors* 37(1): 32–64.

Hoc, J.-M. and Amalberti, R. (2007), 'Cognitive Control Dynamics for Reaching a Satisfactory Performance in Complex Dynamic Situations', *Journal of Cognitive Engineering and Decision Making* 1: 22–55..

Hollnagel, E. (2005), *Accident Models and Accident Analysis*. Available online at: <http://www.ida.liu.se/~eriho/AccidentModels_M.htm>.

Jentsch, F., et al. (1999), 'Who is Flying the Plane Anyway? What Mishaps Tell us About Crew Member Role Assignment and Air Crew Situation Awareness', *Human Factors* 41(1): 1–14.

Marc, J. (2002), *Contribution Individuelles au Fonctionnement sûr du Collectif: Protections Cognitives Contre L'erreur Individuelle et Collective (Le Cas du SAMU)*. Psychology Doctoral Thesis of Psychology, University Paris8.

Marc, J. and Amalberti, R. (2002), 'Contribution Individuelle au Fonctionnement sûr du Collectif l'exemple de la Régulation du SAMU', *Le Travail Humain* 65: 217–42.

Maurino, D.E. Captain (1999), *Human Error in Aviation Maintenance: The Years to Come.* <www.hf.faa.gov/docs/508/maurino15.pdf>, accessed 5 May 2003.
Maurino, D.E., Reason, J., Johnston, A.N. and Lee, R. (1995), *Beyond Aviation Human Factors.* (Hants: Averbury Technical).
Plat, M. (2003), 'Pilots' Understanding Process when Coping with "Automation Surprises"', in Hollnagel, E. (ed.) (2003), *Handbook of Cognitive Task Design* (Mahwah, NJ: LEA).
Plat, M., Rogalski, J. and Amalberti, R. (1998), 'Pilot-Automation Interactions and Cooperation in Highly Automated Cockpits', in Boy, G. et al. (eds.) (1998), *HCIAero'98 Montréal* (CA Editions de l'Ecole Polytechnique de Montréal).
Rasmussen, J. (1997), 'Risk Management in a Dynamic Society. A Modelling Problem', *Safety Science* 27(2/3): 183–213.
Reason, J. (1990), *Human Error* (Cambridge: Cambridge University Press).
Rizzo, A., Ferrante, D. and Bagnara, S, (1987), 'Handling Human Error', in Hoc, J.-M. et al. (eds.) (1987), *Expertise and Technology* (Hillsdale: Lawrence Erlbaum Associates).
Rogalski, J. (1991), 'Distributed Decision Making in Emergency Management: Using a Method as a Framework for Analysing Cooperative Work and as a Decision Aid', in Rasmussen, J. et al. (eds.) (1991), *Distributed Decision Making: Cognitive Models for Cooperative Work* (Chichester: Wiley and Sons).
Rogalski, J. (1996), 'Co-operation Processes in Dynamic Environment Management: Evolution through Training Experienced Pilots in Flying a Highly Automated Aircraft', *Acta Psychologica* 91: 273–95.
Rogalski, J. (1999), 'Decision Making and Dynamic Risk Management', *Cognition, Technology and Work* 1(4): 247–56.
Rouse, W.B., Cannon-Bowers, J.A. and Salas, E. (1992), 'The Role of Mental Models in Team Performance in Complex Systems', *IEEE Transactions on Systems, Man and Cybernetics* 22(6): 1296–1308.
Samurçay, R. and Rogalski, J. (1998), 'Exploitation didactique des situations de simulation', *Le Travail Humain* 61(4): 333–59.
Simon, H. (1982), *Models of Bounded Rationality*, vol. 1 (Cambridge, MA: MIT Press).
Von Cranach, M. et al. (1986), 'The Group as a Self-Active System: Outline of a Theory of Group Action', *European Journal of Social Psychology* 16: 193–229.
Wioland, L. and Amalberti, R. (1999), 'Human Error Management: Towards an Ecological Safety Model. A Case Study in an Air Traffic Control Microworld', in J.-M. Hoc et al. (eds.) (1999), *CSAPC'99 Human-Machine Reliability and Co-operation* (Valenciennes: PUV).
Wioland, L. and Doireau, P. (1995), 'Detection of Human Error by an Outside Observer: A Case Study in Aviation', *CSAPC'95* (Espoo, Finland), 54–62.

PART III

Enhancing Work Practices within Risky Environments

Pascal Béguin

Introduction

Each chapter in this book relates in one way or another to the enhancement of professional practices in risky environments. But this third part asks more specifically: how can efficient and experienced work practices be developed in individual and collective contexts? The two contributors presented here give some indication of the different ways of knowing in risky work environments: during design, or during day-to-day practices. But beyond the specificity of each context, three shared proposals emerge from Part III:

1. Experiential knowing in and from risky work environments;
2. Adapting to the given and the improvised;
3. Enhancing practices individually and collectively.

There is no doubt that the knowledge that is useful is the knowledge applicable to the practical work process at hand, whether for coping with the functional flexibility in air traffic control, or controlling a chemical process. The concern here is not about knowledge in general, which includes much that does not play a role in work practice. The focus here is on ways of knowing that are integral aspects of work practice, and thus essential to support skilled and flexible activities in risky work environments. Employers and trainees frequently complain that vocational courses teach too much inert theory, and industrial psychologists have generally agreed with them. But, traditionally, learning researchers have studied learning as if it were a process contained in the mind of the learner, and have ignored the lived-in world.

An assumption made in this book is that the background and the context in which knowledge is developed during engagement within the risky environments is important. Faverge (1980) defines regulating activities (which include self-regulating activities) as an essential part of action in a risky environment. They consist of bringing into line an unacceptable or problematic state of the system to an acceptable or unproblematic one *before an undesirable event occurs*. This is similar to the definition Owen uses of a near-miss (see Chapter 8). As Owen outlines, a near-miss 'refers to those moments when a trajectory of action is headed towards an error, but is noticed before the error occurs'. In daily risky work practices, errors are avoided most of the time because efficient 'regulating activities' or efficient action takes place in a near-miss context.

A focus on how operators successfully accomplish their work, and mitigate risk, does not only identify strategic dimensions of risky work environments, but also the specific knowledge and associated practices. Thus, Owen (Chapter 8) identifies three dimensions (temporality, complexity and interdependence) that shape the practices of Air Traffic Control and lead to a greater likelihood of near-misses or error. Her study emphasizes that resources are produced by the workers in who create resilience to the threat of mistakes under these circumstances. Béguin (Chapter 7) shows that operators in a chemical process ran the process as close as possible to the low temperature threshold of the product. This is a regulating activity: the workers take the process away from the high threshold temperature, because they are well aware of the high risk of chemical runaway in the upper temperature ranges of the product. But in so doing, they shape another 'daily risk' at lower temperature ranges of the product (the 'crystallization'), which is not taken into account (and is unknown) by the designers of instrumented safety systems.

The second idea common to the following chapters and related to the previous one concerns problematizing the narrow focus of learning theories for the transmission of prior existing knowledge. These theories have a long history in risky work environments. Learning on the floor is not always welcome, particularly in the Taylorist organization. So much so that it appears necessary to Rasmussen (1990) to recall that 'an organization leading to high reliability depends on adaptive self-behaviour. Therefore organizational *learning and adaptation and the implied modification of procedure neither can, nor should be avoided*' (p. 367, editor's emphasis). But in asking what is learned from experience, Rasmussen argued for a better method for analysis of accidents, and a better method to communicate the results to decision-makers. Such an approach is necessary but it remains silent on the creation of new knowledge in practice.

A key message in the chapters in this part of the book is that knowledge and practices constructed by the workers are not given by the designers or the managers. This is particularly the case in dynamic and relatively unpredictable work environments, such as air traffic control (ATC). The controllers must handle the traffic as it happens in real time, with the resources at hand (tools, but also teams). Owen (Chapter 8) argues that, in many aspects, working in ATC is like the activity of 'bricolage'. In this sense, doing is inventive because it is an open-ended process of improvisation with the social, material and experiential resources at hand. So, knowing often arises in a dialectical way: out of efforts to resolve contradictions between what the standard operating procedures are telling the workers to do or what the theory predicts will happen on one hand, and the reality, with its unexpected events that confront them on the other hand. The case presented by Béguin (Chapter 7) highlights such a process, but at the collective level: designers of an alarm learn from the users because of the way they later develop unexpected usage of the artefact.

In each case, experience in risky work practices is beyond the prescribed and provided: the guidelines, the knowledge crystallized in an artefact, or embedded in the system. This distinction seems essential: activity in risky environments cannot be confined to what is given in advance, what is pre-organized or stabilized in artefacts and formal rules and regulations. On the contrary, activity monitors, controls, reorganizes, and in some sense makes use of the given (Béguin and Clot 2004).

A third assumption proposed by the contributors in this Part is that the responsibility for enhancing practices is held collectively as well as individually. Owen (Chapter 8) argues that collective memory is a resource that controllers draw on to access previous experiences of unanticipated disturbances. Thus, the narrative of those events (that controllers call 'war stories') are passed informally between controllers, and illustrate good and bad action in a given context. Bruner (1996) drew attention to the importance of narrative within the framework of cultural psychology. According to him, the building up and the expression of knowledge can evolve along two possible paths: the scientifically logical mode of the model found in natural sciences, and the narrative mode of recounting where the end result is 'to give meaning to experience' (p. 28). War stories appear to match the second mode. They supply the basis for defining the essential points that need to be retained; they promote generalization and enhance learning within the social group. Members of a community create knowledge in the practice of their work, and record their individual perceptions in a structure that constitutes their culture. This leads directly to the idea that organizing for safe environments needs to develop methods of encouraging worker reflectivity. But symmetrically, this also leads to the idea that the development of safe situations depends on the removal, or at least the blurring, of boundaries between different occupational groups. However, in many workplaces a strong division of labour persists. Different interpretations are based on different contextual social positions with inherent differences in possibilities, knowledge, interest and perspective. Béguin (Chapter 7) uses the concept of 'world' to grasp this point, using ideas that come from Cassirer's 'science of culture'. He maintains that in his case, two different social positions – those of the user and those of the designer – constitute two different 'worlds', which never meet. This is a theme also addressed in Part I (in terms of collaboration across boundaries) and in Part II (in terms of the role of common artefacts). So the question is how to design a common world, how to organize a learning system where the strength element of the one complements a weakness elsewhere.

References

Béguin, P., and Clot, Y. (2004), 'Situated Action in the Development of Activity', @ctivités 1.2: 50–63. <http://www.activites.org/v1n2/Béguin.eng.pdf>

Bruner, J. (1996), *The Culture of Education* (Cambridge, MA: Harvard University Press).

Faverge, J.-M. (1980), 'Le travail en tant qu'activité de régulation', *Bulletin de Psychologie* 33.344: 203–206.

Rasmussen, J. (1990), 'Learning from Experience? How? Some Research Issues in Industrial Risk Management', in Leplat, J. et al. (eds.) (1990), *Les facteurs Humains de la Fiabilité dans les Systèmes Complexes* (Toulouse: Octarès).

Chapter 7
When Users and Designers Meet Each Other in the Design Process

Pascal Béguin

Rasmussen (1997) drew attention to the extent of more recent evolutions in risky work situations. As routine tasks are progressively automated, the importance given to activity in situations and user intelligence has constantly grown. Work, as a 'field', has become vast and calls for problem-solving and 'creative improvisation'. As outlined by Rasmussen (2000) these technological evolutions have significant theoretical consequences: efforts are moving from normative models of rational behaviour, through efforts to model observed behaviour regarded as less rational by models of the deviation from behaviour regarded as rational, towards a focus on representing directly the behaviour actually observed, and ultimately towards efforts to model behaviour-generating mechanisms. Vicente (1999) also highlights this evolution: models, which were initially 'normative', became 'descriptive'. However, he argues that a new generation of models is necessary today to apprehend the fact that 'workers finish the design'.

In this chapter, my goal is to discuss the fact that workers finish the design. But my concern is how to build better work systems. So, rather than argue for a new model of behaviour, my focus is to argue for a new model of the design process.

The model known as the 'instrument-mediated activity' (Béguin and Rabardel 2000; Rabardel and Béguin 2005) argues that both designers and workers contribute to design, based on their own competencies. On this basis, it is proposed that understanding and organizing the design process as a mutual learning process is important. In this process a common world is being constructed between workers and designers. The discussion in this chapter presents a case to illustrate the approach: designing an alarm system to guard against chemical runaways in chemical plants.

The Instrument-mediated Activity Approach

The fact that users do not utilize the technical operating system as might be expected, modifying it momentarily or durably, has been pointed out by many authors (see Randell 2003 for a review).

But such a fact leads to various interpretations. One possible interpretation is grounded in the growing collection of workplace studies that emphasize the situated nature of action (for examples in this book, see also Chapters 2 and 5). In work settings, the workers encounter unforeseen events and contingencies linked to 'industrial variability' – for example, systematic deregulation of tools, instability of the matter to be transformed, and so on – and to the fluctuation of their own state, for example, due to tiredness. Thus, tasks and people fluctuate with time, and these fluctuations must be taken into account (Daniellou 1992). Suchman (1987) used the term 'situated action' to generalize this aspect. Whatever the effort put into planning, performance of the action cannot be the mere execution of a plan that fully anticipates the action. One must adjust to the circumstances and address situation-related contingencies, for instance, by acting at the right time and by seizing favourable opportunities. It is this situated approach that was taken up and developed by Vicente (1999), who contended that one must leave workers the possibility to adapt to local circumstances. In risky work, designers' anticipations consist in 'specifying boundaries on action', and allowing workers to finish the design in a 'space of functional action possibilities'. Such an approach is also concerned with better identifying properties that sociotechnical systems should have to allow the workers such spaces for functional action. One challenge is to design 'plastic' or 'flexible' socio-technical systems. They are plastic in the sense that they must incorporate in the activity design sufficient freedom to manoeuvre to deal with situational contingencies (Béguin 2007).

There is a plurality of origins for workers not utilizing the system as might be expected. Situated approaches (briefly discussed below) assume that the environment organizes the subject's conduct and activity (see Béguin and Clot 2004 for discussion). But the situated nature of action not only arises from the dynamic variability of circumstances, as postulated by the proponents of situated action. Driving a vehicle as a job, for example, differs from driving for personal reasons, regardless of the circumstances and the contingencies of the situation. The situated nature of action is also due to activity: workers develop specific instruments but also competencies, as well as subjectively organized forms of action within collectives adapted to their goal and motive.

In the instrument-mediated activity approach (an approach based on activity theory; see Béguin and Rabardel 2000; Rabardel and Béguin 2005), an instrument is seen as a composite entity made up of 'the artefact', in its material and structural aspects, and the subject's schemes. Thus, the mediating instrument has two components: (1) an artefactual one (an artefact, a part of an artefact, or a set of artefacts), which may be material or symbolic, produced by the subject or by others; and (2) one or more associated schemes. Schemes are behaviour organizers; they are the frameworks of actions liable to be repeated.[1] An action scheme is the structured set of action features that can be generalized, that is, that

---

1 Many authors, drawing inspiration from different theoretical frameworks, have worked on the concept of scheme. We will keep here to the elementary definitions of the Piaget school.

make it possible to repeat the same action or to apply it to new contents (Piaget 1970).

Together they act as a mediator between the subjects and the objects of their activity. Indeed, an artefact comprises a set of constraints that the subjects have to manage according to the specificity of each of their actions. Thus, a PC user's activity depends on the interaction constraints specific to the interface. The artefact carries more or less explicit action pre-structuring constraints. So, there are actions related to 'secondary' tasks, that is, tasks related to the management of characteristics and properties specific to the artefact. However, artefacts must not only be analysed as things, but also in the ways in which they mediate usage. Users 'act through the interface', to take up the heuristic expression of Bødker (1989). Schemes consist of wholes deriving their meaning from the global action that aims at operating transformations on the object of activity. The private dimension comes from the particular nature of the historical background and the development of schemes by each of us. Thus, a scheme has characteristics specific to each individual, such as handwriting schemes, that make our handwriting specific and recognizable. The social dimension comes from the fact that schemes develop in the course of a process in which the subject is not isolated. Schemes are shared among practitioners of the same skill and across broader social groups. They are 'shared assets' built up through the creations of individuals or groups (see Chapter 6). They are also the object of more or less formalized transmissions, for example through interpretation of past events transmitted by narrative (see Chapter 8).

Because the instrument is made up of two types of structures, the artefact is far from being finished when the final technical specifications leave the research and design office. Workers seek to exploit the resources available in the environment of their activity and enrol them in the service of action. Authors, such as Scribner (1986), highlighted this dimension, showing that artefacts available in the environment play a decisive role in solving practical problems. For instance, crates and their physical state – empty or full – and their physical positioning – storage organization – help solve concrete counting and calculation problems. To define the role of functions attributed by subjects to objects, Scribner speaks of 'incorporation of the environment into the problem-solving system'. This dimension must be extended to the means of action as a whole: during the activity, and in situations, the user constitutes the artefact (whether physical or symbolic) as an instrument. This is a process that we call an 'instrumental genesis'.

'Catachreses', in the sense intended by Faverge (1970), are the prototype of these processes.[2] The possibility, for a subject, to temporarily associate a wrench with the 'hammering' scheme that is ordinarily associated with a hammer, is a

---

2 The term 'catachresis' is borrowed from linguistics and rhetoric, where it refers to the use of a word in place of another, or in a way which steps outside its normal meaning. By extension, the idea is employed in the field of instrumentation to refer to the use of one artefact in place of another, or to using artefacts to carry out tasks for which they were not designed.

catachresis. The existence of catachreses is a testimony to the subject's creation of means more suited to the ends they are striving to achieve, to the user's construction of instruments to be incorporated into the activity in accordance with current goals. These processes may be relatively elementary, as in using the artefact 'wrench' as a hammer when associated with the hammering scheme. They may also be large-scale processes that develop over a period of years and involve scheme-building as well as transformations of the functions of even the structure of artefacts (see, for example, discussion of CAD in Béguin 2003b). In this case, we speak of instrumental genesis.

Instrumental genesis occurs at both points of the instrumental entity – artefact and schemes – and thus has two dimensions:

1. Instrumentalization, which is artefact-oriented: emergence and evolution of the artefactual components of an instrument – selecting, grouping together, producing, and defining the functions, transformation of the artefact (structure, operation) – which extend the artefact's initial design. We can regard it as a process that enriches the properties of the artefact, because of the attribution of a function to the artefact.
2. Instrumentation, which is subject-oriented. First, schemes assimilate, which means that they are associated with different kinds of artefacts. For example, the hammering scheme, which is usually associated with a hammer, can be momentarily associated with a wrench. Second, schemes are accommodating: they can change when the goal or motive of action changes. Such accommodation is the source of the gradual diversification of uses.

Both of these dimensions are related to the subject. What distinguishes them is the direction in which they occur. In the instrumentation process, the subject develops, while in the instrumentalization process, it is the artefactual part of the instrument that evolves. The two processes contribute jointly to the construction and evolution of the instrument, even if, depending on the situation, one of the processes may be more developed or prominent than the other, or may even be the only one implemented.

Instrumental genesis accounts for a rarely modelled but nonetheless strategic dimension of the user's activity: the constructive dimension of activity (Rabardel and Béguin 2005). The constructive facet of user activity is dependent upon instrumental genesis (its psychological and material components), but it is also dependent upon competencies and conceptualization, and rules in the community. As a general rule, the constructive dimension includes the development, by the subjects, of the conditions and resources of their productive activity. Productive activity is rooted in the subject's intentional commitment to pursuing an objective and attaining goals with the instrument.[3] A dialectical relationship links the productive and constructive dimensions of the activity: failure or resistance at the productive level will lead to new constructions.

The inventiveness and creativity exhibited by users that occurs as they employ and experiment with a new technique is a requirement for its effectiveness and an ontological property of its use. The main point is that users contribute to the design process on grounds and orientations of their own, different from those of institutional designers: both designers and users contribute to design based on their competencies and their own professional role.

Mutual Learning and Interdependence

The approach taken within instrument-mediated activity tries to grasp the inventiveness as it is manifested in the activity of workers confronted with a technique. But the model is not sufficient to guide one's action in the design process. Carroll (1991) points out that the nature of the contribution that may be achieved in the course of design depends largely on the understanding of the available design process. What is the form the design process can take in order to ensure designers' developments are in tune with those of users? How may the designer and user interaction be organized in order to articulate their inventiveness, or more specifically, how may we interrelate and organize the alternation between instrumental genesis and constructive activity by the workers on the one hand, and formal acts of design by the designers on the other?

A Mutual Learning Process

Each designer, by their own activity, learns. This point has been addressed in depth in constructivist epistemology. The famous metaphor of the 'reflexive conversation with the situation' proposed by Schön (1983) is relevant here. According to this author, the design process can be described as an open-ended heuristic during which the designer, striving to reach a goal, projects ideas and knowledge. But then the situation 'replies', and 'surprises' the designer by presenting unexpected resistances. These serve as a learning basis for the designer.

---

3 An analysis of this mediation process shows that the productive part of the user's activity is not only aimed at achieving production objectives (such as quality or quantity). Different forms of mediation can be distinguished. Mediation can be epistemic (aimed at acquiring knowledge about the object that receives the action) or pragmatic (aimed at transforming the object of the action). But in any activity with an instrument, one can also find subject-oriented mediation (reflexive mediation) and collectively-oriented mediation (interpersonal), which serve to regulate the workload and thereby direct and control the action.

However, design is a collective process. The 'reflexive conversation with the situation' often takes place in a 'dialogue' with the object of design. But the object is not alone in talking back; the other actors 'reply' and 'surprise', too. The result of one designer's activity is an hypothesis that will be validated or refuted, or set in motion based on actions and learning performed by another actor involved in the design process. Design is a mutual learning process. Through the hypothesis mediated by intermediary production (scale models, mock-up, prototype), the activity of one designer sets the activity of others in motion. The result of one person's activity constitutes a source for the activity of another.

In such a model, one may consider that each designer, in the course of their activity, and in the framework of their own objectives, undergoes learning processes that are not definitive. That which is learned will be implemented in other people's activity. The initial result will be taken as an object and reworked, which will lead to new learning (Béguin 2003b). Learning carried out within such a process contributes to the changing dynamic of the object being designed.

This point of view is very different from the traditional engineering approach, where design is perceived as a change of state during which a problem must be resolved. In a mutual learning process, design is more of a cyclical process, where the result of one person's activity constitutes a source for the activity of another. A key reason why such an approach in terms of mutual learning is essential is that the inventiveness of both designers and users can be grasped within the same framework. Appropriation of the fruits of one party's labour into the activity of the other party produces something new.

Worlds and Common World

Mutual learning between users and designers is important in participatory design, but many researchers argue that achieving mutual learning between users and designers is a difficult task (see for example, Bjerknes and Bratteteig 1987; Bødker and Grønbæk 1996; Trigg et al. 1991). Bødker and Grønbæk (1996) stress that joint action between users and designers is often seen as creating a new shared activity that is different from that of the designers and that of the users. The 'cooperative design' approach (Kyng 1995) aims to establish a design process wherein both users and designers are participating actively and creatively, based on their differing qualifications. In cooperative prototyping, the multitude of activities (instead of a shared one) implemented around a prototype constitute a place where the future artefact and its use will be developed. We share this viewpoint: design is achieved by separate actors, during which mutual learning is achieved on the basis of the differing qualifications and expertise of the actors.

However, actors are engaged in an interdependent process. A considerable body of research has shown the existence of contradictory organization and/or technical logic (De Terssac 1992; Leplat 2000), particularly disturbing in risky work (Rasmussen 1997). The heterogeneous views had to be articulated and adapted to form a joint work. This is the very purpose of 'project management' (Charue-Duboc and Midler 1998). Furthermore, designers may not draw any benefits from the constructive activities of users. The reverse is also true. In the instrument-mediated activity model, catachreses or instrumental genesis reflect the production of a means of action by workers, with their own 'point of view'. However, instrumental genesis is often rightly regarded as deviant use of an artefact with respect to the functions intended or imagined by the designers. Such deviations cause problems in the dangerous situations they may create. Distorted uses may not be compatible with the underlying principles of the technical process, and finally with the point of view of the designers. The different points of view must be articulated during the design process, for constructing something new. In fact, the question is: what is the object of the learning process? 'Point of view' is a vague term that needs conceptualization.

One is tempted here to define the 'point of view' as a 'world', that is, as a way of grasping a situation (that is neither entirely objective nor entirely subjective) whose function is to conceptually organize reality and orient action. The notion of world, which suggests that there exist different possible and acceptable worlds for understanding the same situation of action[4] (Goodman 1978), has been widely examined in the literature, sometimes in diverging ways. In design, it is most often employed to account for the social and cognitive 'features' of a given specialist (for example, a methods engineer). Bucciarelli (1994) showed, for example, that two designers faced with the same artefact mobilize different 'object worlds'. Each one utilizes a jargon, concepts, and even values, all connected into a system that makes the object take on different forms. This approach is interesting insofar as it shows that the conceptual, axiological and praxeological backgrounds of an actor in the design process form a system with the object they are specifying or developing.

But the different 'worlds' cannot be left in a disjointed state. The need for these differences to be articulated, to promote the establishment of a 'common world', is a concept used by Cassirer (1942). According to Cassirer, the primary (in the genetic sense) common world is language. Language creates a link between the individual and all else; it gives the individual access to the supra-individual, to articulate and organize the contents of individual experiences and intuitions in a specific and stable mode. But this shared reference is not monolithic. Take the example of the term 'work'. This notion has been the object of a lengthy historical construction among the working classes, through which its members interpret situations and grant meanings to work experiences. But 'work', as an historical construct, is constantly undergoing a local redefinition process. Access to a common world, then, is not a passive reception process but a highly creative act, one where 'each individual appropriates for himself, and works, through this appropriation and thanks to it, towards its maintenance and its renewal' (Cassirer 1942, 91; my

---

4   The distinction made in Russian psychologists between 'objekt' and 'premyet' is relevant here. Premyet corresponds to subjects' subjective grasp of reality.

translation). A common world makes it possible to distinguish, articulate and organize the contents of individual experiences and world in a specific and stable mode. So, a common world is not a monolithic entity, but a system of worlds to articulate, a learning system with many entries, whose contradictions impel development.

Constructing Common Worlds Through Mutual Learning

How does building a common world occur, through mutual learning between users and designers? And how may such learning be understood in the instrument-mediated activity model?

Introducing Bakhtin's ideas about 'dialogicality' as a way to expand the Vygotskian approach, Wertsch (1998) (who considers language to be a cultural tool, and speech to be a form of mediator) stresses the importance of appropriation of utterances during production. The term 'appropriation', borrowed from Bakhtin (prisvoenie) refers to a process where someone takes something that belongs to others, and makes it his or her own. As Wertsch outlined, producing an utterance inherently involves appropriating the words of others, and making them, at least in part, one's own.

> Because words are half-ours and half-someone else's (Bakhtin 1981, 345), one is invited to take the internal word as a 'thinking device', or as a starting point for a response that may incorporate and change the form or meaning of what was originally said. (Wertsch 1998, 67)

My goal here is not to argue that there is no difference between artefacts and words.[5] But it is contended that dialogicality is not a specific dimension of language. In the design process, exchanges between actors are carried out through the mediation of many intermediary productions (temporary outcomes of the design activity, not just through words), whose use in action reshapes, enriches or shifts the characteristics of the object currently being designed. But the process during which the result of a designer's activity is brought back into play in the activity of a user is also a dialogical process. The consideration that it is too restrictive to limit dialogicality to language exchanges allows one to grant a status to a critical point in the instrument-mediated approach: the instrumental genesis. Instrumental genesis is a way of conceptualizing appropriation processes (Béguin and Rabardel 2000; Rabardel and Béguin 2005. The 'someone else's half' (the artefact) is associated with 'one's own half' (the scheme) to bring about instrumental genesis.

---

5 Even if we agree with others (for example, Cole 1995) that the distinction made by Vygotski between two classes of instruments (technical instruments and psychological instruments) is not always relevant.

During this process, a user's action reshapes or enriches the artefact. But designers need to learn more. Designers and users contribute to design based on their diversity, and designers have their own qualifications and expertise.

Due to constructive activity, appropriation is revealing a developmental process during which a user either adapts the conditions and resources available in the situation to their own world (as is the case during instrumentalization), or adapts its own world to the situation (as is the case during instrumentation). Appropriation constitutes an opportunity for revealing worlds of the users: schemes associated with artefacts, competencies and conceptualization, goal and motive.

In this light, an approach like 'rapid system prototyping in collaboration with users' (Shapiro 1994, 425) probably only grants a minor role to the users' activity. A detour, consisting in analysing work in which instrumental genesis manifests itself, is necessary.

First, mutual learning is achieved within a community, where the diverging views that surface may lead to mutual learning, but also to the exclusion or compliance of certain actors. The possibility of excluding certain actors is even more likely during joint action among users and designers, as shown by Gärtner and Wagner (1996), for whom 'an important part of agenda setting is to create legitimacy' (p. 211). Cicourel (1994) showed how a shift in sociological thinking can be achieved by placing knowledge production at the core of the analysis of the social organization. The latter is less related to an established order than to a negotiated order that sets in step by step as new transactions take place. We think that work analysis can play an essential role in this context. It is capable of eliciting knowledge that highlights the nature of the problems operators encounter on the job (Wisner 1995) and of showing that their actions, skills and know-how are objectively adapted to the situation. In other words, work analysis can help operators to be treated as worthy contributors. In this respect, work analysis is a prerequisite to setting up a mutual learning process.

Second, we need to 'look for invisible work' (Blomberg et al. 1996). But in addition to being difficult to see, not all constituents of activities with instruments are directly verbalizable. This is particularly true of schemes in our approach, which have to be grasped. Action is known to mobilize 'incorporated knowledge' which is difficult to verbalize (Leplat 2000). This dimension also applies to schemes, which are supported by 'operating concepts' (Vergnaux 1996) or 'pragmatic concepts' (Samurçay and Pastré 1995) that cannot be put into words by their users. A large part of the activities at work, even in very modern settings, may fall into this category. Work analysis in this case has a twofold objective. It is aimed first of all at informing us of users' worlds. But it is also a condition for user participation in the transformation of their work situations. This is because the method equips users; it supplies them with the conceptualization they need to talk about their own activity.

A work analysis enables an 'objectification', that is, making the constructive activity of users available and legitimized within the mutual learning community. It is not until this objectivation task has been completed that constructing a common world is possible. From a methodological point of view, it consists of arriving at a collective interpretation of events: 'We observed something. What should we do, what lessons can be drawn from it, what decisions should we make accordingly?' Bruner (1996) calls this collective interpretation a 'record'. The record supplies the basis for defining the essential points that need to be retained; it promotes generalization, capitalization and memorization; it supports and facilitates learning.

Designing a Safety Instrument System

We will illustrate this approach during the design of an alarm. The project was launched following an inquiry conducted in a chemical plant where an operator had been killed due to an explosion caused by a 'chemical runaway'. Five operators were on-site at the time of the accident. They all said they had realized that something was wrong two hours before the explosion, but did not attempt to leave the premises until a few seconds before it occurred. There are several reasons why the operators stayed on-site too long, including the desire to 'recover' production and avoid destruction of the installation. But the main problem was their difficulty in identifying the time remaining before the explosion. This point was confirmed by other analyses (accidents like this are not rare in chemical plants, especially in small or medium-sized companies). Chemical runaways are one of the major causes of human death in sites classified as SEVESO. The decision was thus made to develop an instrumented safety system[6] (Béguin 2003a) in order to prevent the occurrence of this serious type of accident. An algorithm to detect the critical moment of chemical runaway (the explosion) was developed, and then tested in an experimental situation (large-scale test).[7]

Once the algorithm was developed, the engineers contacted us to ask questions about 'the appropriation of the device' (as they call it). They wondered, for example, whether the device would lead operators to push the reaction process to its limits for the purposes of production. After having explained our position, we proposed to install a prototype of the device (beta version) at a pilot site interested in having such an alarm at their disposal.

---

6 Three types of process control devices are defined by norm NE 31, depending on the extent of the risk: control systems, which assist operations; monitoring systems, used to detect events; and instrumented safety systems, which are specifically devoted to deteriorated situations.

7 The algorithm detected about 60 per cent of the accidents described in the database developed by Marss et al. (1989). Despite the lack of specific studies on this topic, the algorithm proved capable of predicting the time-to-maximum-rate (TMR), based on the product's temperature curve, with a reasonably good error level.

The alarm and the pilot site

The prototype consisted of an anti-deflagration box, two alarms (visual and auditory), and two digital displays. One display gave the temperature and the other, the 'time-to maximum-rate' (TMR). The TMR predicts the time remaining before a chemical runaway, on the basis of the product's dynamics. From the designer's viewpoint, the purpose of the device is to structure the operators' actions: an initial visual alarm would indicate the need to take preventive action, and then a sound alarm would be a warning to leave the premises.[8]

The device was installed on a catalyst production unit that makes synthetic optical lenses. Five operators put out two 'loads' of the product daily. A 'load' is several tons of product from the catalyser, where it is synthesized in two glass reactors.

The operators work right beside the reactors, with no 'control room' between them and this highly explosive product. The product can take on three states:

1. A gaseous state. The synthesis of the catalyser is exothermic: the temperature of the product can rise on its own and release gaseous by-products. If the exothermic process 'takes off', the build-up of gas can be so great that there is an explosion. This is what is called 'chemical runaway'.
2. A solid state. To avoid chemical runaway at the pilot site, the catalyser is cooled during synthesis by glycolated water circulating in coils placed in the reactors. Thermal homogeneity is ensured by agitators. But over-cooling of the product can lead to an unstable 'super-undercooling' phase. The product then solidifies: this is what the operators call 'crystallization'.
3. A liquid state. Between the solid 'super-undercooling' state and the gaseous state in chemical runaway, the product is liquid.

The Prototype in the Users' World

The results obtained in the weeks following the introduction of the artefact showed that the operators looked at the alarm increasingly often. During the first session, they consulted the interface only 1.7 per cent of the total process control time, whereas they did so 31.5 per cent of the time by the end of the sixth session[9] (see Table 7.1). However, a finer analysis of information intake (gaze direction; for additional methodological details see Guerin et al. 1997) pointed out the paradoxical nature of this 'appropriation'. The device merely replaced

---

8 Although simple in form and function, this safety device is nonetheless based on complex principles and is highly innovative. No comparable device was available on the market at the onset of the study.

9 The artefact was put into operation one day a week, in our presence. A total of 11 sessions were held.

previously available thermometers (see Table 7.1). In other words, operators read the temperature information off the artefact, but not the TMR, even though this was the main advantage offered by the new system.

This is a case of instrumental genesis, and more specifically, of instrumentalization (see discussion below): the users assimilated the artefact into their existing operating schemes by assigning it a function that differed from the one initially planned by the designers. In the present case, the operators gave the artefact the function of a thermometer, and not the function of an alarm. Instrumental genesis originates in a 'world of cold', very different from the engineers' 'world of hot'.

The activity analysis showed that the operators ran the process 'as close as possible to crystallization', based on the low temperature threshold rather than on the high threshold (the chemical runaway point). This strategy is rooted in two facts:

First, obviously, operators are well aware of the high risk of chemical runaway in the upper temperature ranges of the product. Keeping the process running close to the crystallization state lowers this risk.

Second, production experience has shown that the colder the product during synthesis, the higher the quality of the resulting catalyzer (from the physico-chemical standpoint). In controlling the process, then, operators maintain the product as close as possible to the lower temperature threshold for production purposes.

However, with this low-temperature strategy, there is a risk of crystallization. Although less dangerous than chemical runaway, crystallization is an 'ongoing risk', a serious incident that must be avoided. It leads to solidification of the product, which may cause equipment breakage (especially of the agitators, which help prevent solidification and promote product cooling). In addition, when crystallization does occur, the product must be allowed to warm up in one way or another, with all the risks the instability of this polymer entails.

Controlling the process close to the lower temperature threshold is a difficult task which has corresponding schemes and requires skills and conceptualizations. The operators mobilized a 'world of cold' composed of cooling systems,

Table 7.1 Comparative evolution of operators' gazes (in percentage of total session observation time) at the prototype and at previously available thermometers during the first, third, and sixth working hypothesis sessions

Operator's Gaze	First session	Third session	Sixth session
At prototype	1.7%	8.1%	31.5%
At previously available thermometers	31.3%	24.5%	4.2%
Other (at product, co-worker, etc.)	67%	67.4%	64.3%

'beginnings of crystallization' and 'crystals', a world constructed for taking action during the process, and whose frame of reference was crystallization. And yet this world is not monolithic. Two versions of the same world were identified, despite the small number of operators.

In one strategy, it is deemed important to avoid propagation of the 'beginnings of crystallization', that is, the formation of crystals in the vicinity of the cooling system. In this strategy, the agitators are stopped as soon as these 'beginnings of crystallization' appear. This strategy rests on a 'propagation theory' regarding the onset of crystallization.

In another strategy, it is deemed on the contrary that the agitators must be kept running, especially after the last bout of cold. In this approach, it is believed that agitation optimizes thermal exchange between the cooling system and the product. This strategy can be said to rest on a sort of 'thermal equilibrium theory'.

The 'propagation theory' and the 'thermal equilibrium' theory elicited during the activity analysis were not based on scientifically validated knowledge, but were conceptualizations about what action should be taken to avoid the 'ongoing risk' of crystallization.

Whatever the different versions, it is the world of cold that gives sense to instrumental genesis. The devices available to the operators before the prototype was introduced were inadequate in this respect. There were sensors that indicated the temperature to the nearest half degree, whereas controlling the process close to crystallization requires a precision level of about one tenth of a degree. As a result, the operators had to make estimations. Such estimates were no longer needed with the alarm, which displayed the temperature in tenths of a degree, thereby allowing for greater process-control precision.

The users' 'world of cold' is, however, very different from the engineers' 'world of hot'; engineers are living in a gaseous and explosive world grasped through in vitro study, followed by minute observations loaded with calculations and formulas. It is a world constructed for taking action during that same process via the artefact, but whose frame of reference was hot, and chemical runaways.

Redefining the Project

This instrumentalization process taught the designers something, and with this in hand, they developed a second artefact. An analogical temperature display with memory was added to the artefact along with the digital display. An analogical temperature curve would facilitate the interpretation of a trend in the thermal kinetics of the product, and would act as a preventive variable of chemical runaways.

The engineers were nevertheless in an uncomfortable position. The instrumentalization was a sign of failure, for two reasons. The use that the operators made of the alarm violated Norm NE 31, which stipulates that 'regular control systems', 'monitoring systems' and 'instrumented safety systems' should be separate. The way that the operators used the new artefact, initially designed to be an instrumented safety system, made it into a monitoring system. Furthermore, while the device enabled the operators to prevent crystallization, what role did it play relative to the main risk? The artefact, whether anyone liked it or not, had turned into a mere thermometer. Given the research and development costs, it was a very expensive thermometer indeed!

In order to get around this obstacle, we suggested 'reflective backtracking' sessions where the plant foreman, the operators, the engineers and the ergonomists would go back over and think about the study results. The goal was to establish a record, in the sense of the term Bruner (1996) used and defined above. The record led to a redefinition of the purpose of the project, which can be described in three steps.

The first step consisted in weighing up the different worlds. The introduction of the alarm had put two forms of expertise face to face. The first pertained to crystallization, which was a lesser but still ongoing risk. The operators had developed their skills in this realm. They were crystallization experts. The second type of expertise pertained to the risk of explosion. The artefact, or more exactly, the theoretical knowledge upon which it was based and embodied, was the outcome of chemical runaway expertise.

The second step involved defining the interdependency of these two points of view. The point of view based on the major risk of chemical runaway, which served as the frame of reference for the engineers, did not contribute to controlling this particular process, with all its various characteristics (size of reactors, load variability which depends on the outside temperature, among others). Nor did it help to cope with the daily risk. Reciprocally, crystallization, which was the frame of reference for the users, was also insufficient. Chemical runaway could appear following poorly controlled super-undercooling, or simply following equipment breakage. If these events were to happen, the operators would have to 'cope with the unknown' because the opportunity for building knowledge of chemical runaways on the job was non-existent: one must produce without risk. They had to function under uncertainty, as in all risky situations: the more efficient the operators were at protecting themselves, the less opportunity they had to learn!

The third step involved establishing a new orientation for the project. While there was agreement upon the goal to attain (to lower the risk of death), the range of possible solutions was only very partially dependent upon the design of the alarm. Not only was crystallization expertise a necessity that had to be instrumented – the second version of the artefact fulfilled that function – but at the same time, the workers needed to learn from the designers' world.

Workers' Activity in the Designers' World

From our standpoint, the record had shown that the common world was incompletely constructed. So far, the designers had encountered the users' world.

To complete the learning process, the next step was to confront the worker activity with the designers' world.

The characteristics of the artefact made it possible for the user to experience, in action, the concrete conditions of a chemical runaway. The artefact operated on the basis of an algorithm that modelled the thermal kinetics of the product until chemical runaway occurred. The artefact thus enabled one to simulate a runaway. This involved developing a new version of the alarm and devising scenarios so that the operators could experiment in action with how they would have to act in the case of a chemical runaway.

Of course, the artefact had to be operated without the product, so it was necessary to develop a new version of the prototype. During the working hypothesis sessions, the polymer was replaced by an inert liquid.

Scenarios were defined on the basis of the company's safety procedures. These procedures were seen as methods geared to respond to well-defined situations which are all the more important as the situation is rare[10] as is often the case here. Three scenarios were defined, corresponding to the different ways of inhibiting the process (destruction of the product or chemical inhibition of the reaction).

A simulation was run with the operators and other persons in charge (as stipulated in the procedures), along with the engineers.

For each scenario, a record was produced by ourselves and the engineers. The records provided the opportunity to transmit knowledge to the operators about chemical runaways. However, from our point of view, the most important fact was not the enhancement of the operators' conceptualizations and skills. In experimenting with the concrete conditions of product destruction, it became apparent in two of the scenarios that it was impossible for the operators to prevent chemical runaway! The first scenario pointed out the need to change the organization's means of action. Indeed, there were not enough operators on-site to handle product destruction. The second scenario pointed out the need to modify the working conditions rather than the means of action. In this second scenario, it was discovered that the architectural characteristics of the production room (in particular, the state of the floor) hindered proper action.

Conclusion

The results just presented pointed out the importance of a dialogical process – from workers to designers versus from designers to workers – during which intermediary versions of the artefact being designed serve as a vector for learning. The introduction of the first version of the artefact led to its instrumentalization by the operators. This instrumental genesis, 'objectified' through a work analysis, led to the development of a second version of the prototype by the designers. After this first phase, the project was reoriented in order to allow the users to experience the

---

10   Rogalski (1996); personal communication.

concrete conditions of a chemical runaway. Once implemented, this phase showed that the users would fail under these circumstances, due to the organizational and architectural conditions of the site. This launched a final cycle of design in which we did not participate: organizational and architectural modifications were made a few weeks after our project had ended. An additional operator was hired, and the floor of the production room was modified. When sufficiently long and sufficiently objectified, activity exchanges like these substantially modify the object being designed. Initially pragmatic in nature the device took on its final status of a subject-oriented instrument with a quasi-didactical purpose.

But it seems impossible to fully account for the process without postulating different levels of learning: at the level of the artefact, and at the level of the world. We have seen that to begin with, the designers 'learned' from user appropriation. Based on this, they designed a second version of the alarm. But this was only a first 'learning' level. It was a prerequisite to the second level, a much more important one since it provoked a reorientation of the project. In other words, in the first step the designers draw a new version of the alarm out of the users' appropriation of the artefact; in the second step, the designers gain a new understanding of their own activity from the users' activity.

This suggests two levels during activity exchanges: learning and development. This is an important distinction and has been the subject of many debates which cannot be reported here. Let us simply stress the importance of what Bateson (1972) calls a 'double bind'. Bateson worked out a well-known, complex hierarchy of learning processes based on a 'hierarchic classification of the types of errors which are to be corrected in the various learning processes' (Bateson 1972, 287). A central aspect of Bateson's theory is that it insists on the role of the subject's inner contradictions. In a double-bind situation, the individual receives two messages, or commands, which negate each other. It is these inner contradictions at one level that generate learning at a higher level. During the activity exchanges, the designers themselves expressed this double bind: 'If there was appropriation, it wouldn't work, and if there wasn't, it was useless'. Resolving this dilemma is what triggered the reorientation of the project, which was then attributed a new meaning. But the double bind appeared as a highly critical phase, one of discouragement in the designers and the desire to end the study. From a methodological standpoint, this leads us to pay careful attention to what we have called 'objectification' and 'record'. The idea is to seek a way out of the contradictions, and even more broadly, a way of producing something new. The outcome of this process is uncertain. It also encourages us to be attentive to conditions where users can be true actors, where they will be able to have an impact on the choices made during the search for a new way of doing design.

References

Bakhtin, M.M. (1981), *The Dialogic Imagination: Four Essays* (Texas: University of Texas Press)

Bateson, G. (1972), *Steps to an Ecology of Mind* (New York: Ballantine Books).
Béguin, P. (2003a), 'Conception et développement conjoint des situations et des Sujets' <http://www.cnam.fr/ergonomie/labo/equipe/beguin/articles_pb/Beguin2003.pdf>
Béguin, P. (2003b), 'Design as a Mutual Learning Process between Users and Designers', *Interacting with Computers* 15.5: 709–730.
Béguin, P. (2007), 'Taking Activity into Account During the Design Process', *@ctivités* 4.2: 115–21 <http://www.activites.org/v4n2/v4n2.pdf>
Béguin, P., and Clot, Y. (2004), 'Situated Action in the Development of Activity', *@ctivités* 1.2: 27–49.
Béguin, P. and Rabardel, P. (2000), 'Designing for Instrument Mediated Activity', *Scandinavian Journal of Information Systems* 12: 173–90.
Bjerknes, G. and Bratteteig, T. (1987), 'Florence in Wonderland', in Bjerknes, G. et al. (eds.) (1987), *Computers and Democracy – A Scandinavian Challenge* (Aldershot: Avebury).
Blomberg, J., Suchman, L. and Trigg, R.H. (1996), 'Reflection on a Work Oriented Design Project', *Human-Computer Interaction* 11.3: 237–66.
Bødker, S. (1989), 'A Human Activity Approach to User Interfaces', *Human Computer Interaction* 4: 171–95.
Bødker, S. and Grønbæk, K. (1996), 'Users and Designers in Mutual Activity: An Analysis of Co-operative Activities in Systems Design', in Engeström, Y. and Middleton, D. (eds.) (1996), *Cognition and Communication at Work* (Cambridge: Cambridge University Press).
Bruner, J. (1996), *The Culture of Education* (Cambridge, MA: Harvard University Press).
Bucciarelli, L.L. (1994), *Designing Engineers* (Cambridge: MIT Press).
Carroll, J.M. (1991), 'The Little Kittle Little House', in Carroll, J. (ed.) (1991), *Designing Interaction: Psychology at the Human Computer Interface* (Cambridge: Cambridge University Press).
Carroll, J.M. (ed.) (1995), *Scenario-Based Design: Envisioning Work and Technology in System Development* (Chichester: John Wiley & Sons).
Cassirer, E. (1942/1991), *Logique des Sciences de la Culture (Logic of Cultural Sciences)*. (Paris: PUF).
Charue-Duboc, F. and Midler, C. (1998), 'Beyond Advanced Project Management: Renewing Engineering Practices and Organizations', in Lundin, R.A. and Midler, C. (eds.) (1998), *Projects as Arenas for Renewal and Learning Processes* (The Netherlands: Kluwer).
Cicourel, A.V. (1994), 'La connaissance distribuée dans le diagnostic médical', *Sociologie du Travail* 36.4: 427–50.
Cole, M. (1995), 'Socio-cultural-historical Psychology: Some General Remarks and a Proposal for a New Kind of Cultural-Genetic Methodology', in Wertsch, J.V. et al. (eds.) (1995), *Sociocultural Studies of Mind* (Cambridge: Cambridge University Press).

Daniellou, F. (1992), *Le Statut de la Pratique et des Connaissances dans L'intervention Ergonomique de Conception (Status of Practice and Knowledge During Design Ergonomics Intervention)*. Thèse D'habilitation à Diriger des Recherches. (Université de Toulouse-Le Mirail).

De Terssac, G. (1992), *L'Autonomie dans le Travail* (Paris: PUF).

Faverge, J.M. (1970), 'L'homme agent d'infiabilité et de fiabilité du processus industriel', *Ergonomics* 13.3: 301–327.

Gärtner, J. and Wagner, I. (1996), 'Mapping Actors and Agendas: Political Frameworks of Systems Design and Participation', *Human-Computer Interaction* 11.3: 187–214.

Goodman, N. (1978), *Ways of Worldmaking* (Indianapolis: Hackett Publishing Company).

Guérin, F., Laville, A., Daniellou, F., Duraffourg, J. and Kerguelen A. (1997), *Comprendre le Travail pour le Transformer, La Pratique De l'Ergonomie*. 2nd edition (Lyon: ANACT).

Kyng, M. (1995), 'Creating Contexts for Design', in Carroll, J.M. (ed.) (1995), *Designing Interaction: Psychology at the Human Computer Interface* (Cambridge: Cambridge University Press).

Leplat, J. (2000), *L'Analyse Psychologique du Travail en Ergonomie (Psychological Analysis of Work in Ergonomics)* (Toulouse: Octarès)

Marss, G. P., Lees, F.P., Barton, J. and Scilly, N. (1989), 'Overpressure Protection of Batch Chemical Reactors', *Chemical Engineering Research and Design* 67: 381–406.

Piaget, J. (1970), *Biologie et Connaissance (Biology and Knowledge)* (Paris: PUF).

Rabardel, P., and Béguin, P. (2005), 'Instrument Mediated Activity: From Subject Development to Anthropocentric Design', *Theoretical Issues in Ergonomics Science* 6.5: 429–61.

Randell, R. (2003), 'User Customisation of Medical Devices: The Reality and the Possibilities', *Cognition, Technology and Work* 5.3: 163–70.

Rasmussen, J. (1997), 'Risk Management in a Dynamic Society: A Modeling Problem', *Safety Science* 27.2/3: 183–213.

Rasmussen, J. (2000), 'Human Factors in a Dynamic Information Society: Where are we Heading?', *Ergonomics* 43.7: 869–79.

Samurçay, R. and Pastré, P. (1995), 'La conceptualisation des situations de travail dans la formation des compétences (Conceptualization of Work Situation During the Development of Expertise)', *Education Permanente* 123: 13–32.

Schön, D. (1983), *The Reflective Practitioner: How Professionals Think in Action* (HarperCollins USA).

Scribner, S. (1986), 'Thinking in Action: Some Characteristics of Practical Thought', in Sternberg, R and Wagner, J. (eds.) (1986), *Practical Intelligence: Nature and Origins of Competence in the Everyday World* (Cambridge: Cambridge University Press).

Shapiro, D. (1994), 'The Limits of Ethnography: Combining Social Sciences for CSCW', *Proceedings of the CSCW'94 Conference on Computer Supported Cooperative Work* (Chapel Hill, NC: ACM), 417–28.

Suchman L. (1987), *Plans and Situated Actions* (Cambridge: Cambridge University Press).

Trigg, R.H. et al. (1991), 'A Video-Based Analysis of the Co-operative Prototyping Process', *Scandinavian Journal of Information Systems* 3: 63–86.

Vergnaux, G. (1996), 'Au Fond de l'Action la Conceptualisation', in Barbier, J.M. (ed.), *Savoirs Théoriques et Savoirs d'action* (Paris: PUF).

Vicente, K.J. (1999), *Cognitive Work Analysis: Toward Safe Productive and Healthy Computer-Based Works* (Mahwah, NJ: Lawrence Erlbaum Associates).

Wertsch, J.V. (1998), *Mind as Action* (Oxford: Oxford University Press).

Wisner, A. (1995), 'Understanding Problem Building: Ergonomic Work Analysis', *Ergonomics* 38.8: 1542–83.

Chapter 8
Near-misses and Mistakes in Risky Work: An Exploration of Work Practices in High-3 Environments

Christine Owen

Is work risky because of the fallibility of the humans involved or are there particular working conditions and task demands that are inherently error ridden and vulnerable to failure, and where safety is only accomplished through the skilful practices of humans in mitigating the consequences of such activity? This question has been at the heart of investigations into high-reliability work environments for decades. Of course, both are implicated and interact, particularly in the practice of work in complex and interdependent contexts.

High-3 work environments create, by definition, risky work practices. The term 'High-3 work' describes practices that have high-technology, high-intensity and high-reliability characteristics. High-technology practices are those that have a strong reliance on information-communications technologies that can also include work that is either automated or semi-automated and thus mediated by computers. High-intensity work occurs in contexts that have time pressures and often with a strong sense of urgency. This is because the dynamic 'real-time' nature of the work contains elements that, once initiated, cannot be stopped and must be managed as events evolve. Within such workplaces, the focus is also on how to create high-reliability (or 'resilient' performances – see Chapters 1 and 2 for a discussion), because an error can potentially lead to unacceptable consequences.

High-3 work practices are observable in virtually all industry sectors, although more developed examples are to be found within transportation, emergency services, health services, certain manufacturing (for example, chemical industries), the military and police, and key elements of the finance sector.

These work environments are more vulnerable to risk, in part, because there is a greater scope for error in work practices that involve high-intensity operations. The pressure felt from working in a real-time environment characterized by short timeframes often creates a greater chance of mistakes. The complexity of the work arises from managing multiple possibilities of outcomes which also increases the scope of error in judgement and coordination. Likewise, work with an element of high-reliability is, by definition, one where the consequences of error are substantial. Moreover, as systems become more tightly coupled (Perrow 1999) in their integration, the potential effects of an error within these systems increase in consequence and magnitude (Hollnagel 2006; Perrow 1999; Zuboff 1988).

My focus in this chapter is to provide a detailed analysis of work practices in one High-3 work environment: that of air traffic control (ATC) in order to map the features of this domain of work. In so doing, the focus here is to address what Barley and Kunda (2001, 76) call the need to 'bring work back in' to theorizing. These authors contend that studies of work practice have more recently fallen out of favour, being replaced by organizational theorizing. In this shift, they argue, we have lost focus on what it is that people actually do. Barley and Kunda call for a refocusing on detailed studies of work practice to revitalize theoretical understandings, 'not of *organizations* but of *organizing*' (p. 85; emphasis added). Such an approach, they claim, 'may enable researchers to break new conceptual ground, resolve existing theoretical puzzles, and even revitalize older concepts by which we envision the nature of organizing' (Barley and Kunda 2001, 86). My aim here is to take up this challenge and to look for new insights into the practices of undertaking risky work and how such work can best be supported.

Thus, my purpose is to suggest that, rather than attempting to create error-free environments (which may indeed be utopian), we would be more realistic and thus more helpful to actors involved in High-3 work domains (and their stakeholders) if we could identify the individual, group, systemic and technological resources that assist operators to manage near-misses as and when they occur and to develop systemic continuous learning processes to create better workplaces for the future. This is not to dismiss approaches aimed at error elimination but rather to recognize their limitations, and, through developing an understanding of what really happens in everyday work practice, build better resources to support individuals, teams and systems.

Near-misses refer to those moments when a trajectory of action is headed towards an error, but is noticed before the error occurs in such a way that error can be corrected. Averting an error can occur with or without the awareness of the actors involved. An error or mistake occurs when planned actions (which might include no action) fail to achieve their desired consequences.

This discussion commences with an overview of some of the more recent approaches to understanding near-misses and mistakes in other High-3 work environments before proposing an alternative framework based on work practice. Data from two empirical studies will then be used to illustrate the ways in which a more nuanced understanding of work practice can shed light on the elements that create the possibility of the occurrence of near-misses and mistakes.

Examining Performance in High-3 Practices: The Case of the Aviation Industry

The notion of risk-taking and the aviation industry have historically been synonymous. The aviation industry was born from the risks of individuals attempting to do something previously not possible. The stereotypical images of success, characterized by the aircraft pilot, 'a single, stalwart individual, white scarf trailing, braving the elements in an open cockpit' (Helmreich and Foushee 1993, 4), have led to acceptance – indeed celebration – of norms associated with 'independence, machismo, bravery, and calmness under stress' (Helmreich and Foushee 1993, 4). The history of these experiences led to assumptions that in addition to technical competence, successful performance depended on having these personality traits. This view began to change with the arrival of the jet age when investigations into aircraft accidents began drawing conclusions that 'pilot error' was more likely to reflect failures in team communication and coordination than deficiencies in an individual's technical skills (Murphy 1980). In the aviation industry, the technology involved has become more reliable and this has resulted in an increasing proportion of errors being attributed to humans operating in the system (Hartel and Hartel 1995).

In many of these cases, the information needed at the time was available somewhere within the organizational system, but had either not been passed on, or had been passed on incorrectly (Hartel and Hartel 1995). This has led to a call to develop 'requisite variety' (Weick 1987) through a 'culture of conscious inquiry' (Westrum and Adamski 1999). According to Weick (2001), 'requisite variety' is needed in complex systems because the complexity (variety) that occurs within those systems exceeds the individual capability of the people required to manage them. 'When people have less variety than is requisite to cope with the system, they miss important information, their diagnoses are incomplete, and their remedies are short-sighted and can magnify rather than reduce a problem' (Weick 2001, 331). Requisite variety can be enhanced through the skills of conscious inquiry by enabling people to pool the capability they have. It is contended that requisite variety is drawn from a compilation of individual operators, effective teamwork and systemic processes and technologies that make relevant information available. The development of 'learningful work environments' by designing cultures and structures that enhance requisite variety was the focus of an earlier study and upon which some of the empirical data discussed here is drawn (Owen 1999). Through a close examination of work practices, it is hoped that a better understanding will be gained of when near-misses and mistakes are more likely so that this understanding can be developed to create more learningful work environments.

Examining air traffic control work practice

The data reported here are drawn from two sources. The first was a five-year qualitative study that involved observations and interviews with a stratified sample of 100 air traffic controllers across three ATC centres in Australia (Owen 1999). The second subsequent study, also employing observation and interview strategies, has examined the impacts of changing technologies on, within and among team performance. The epistemological orientation in both studies was a phenomenological one aiming to better understand the lived experiences of people in their work.

In both studies, observations and interviews were also conducted across the three main operational functions of ATC (Tower, Approach/Departures and Enroute Control). The methods used and the process of data analysis has been described elsewhere (Owen 1999; 2001). In collecting the data, observations and interviews were conducted with controllers who had varying levels of expertise. This included trainees as they learned the job of air traffic control, and more experienced air traffic controllers. In some cases controllers were interviewed on multiple occasions over an 18-month period. The following outline of ATC work is provided to show how the work is organized.

Organization of ATC Work Practice

Within Australia, there are approximately 1,200 air traffic controllers, managing air traffic in and out of 29 regional, national and international airports. The volume of airspace controlled by ATC in Australia covers approximately 10 per cent of the world's surface, or 15.6 million square nautical miles (MacPhee 1992). Most of the ATC work occurs in two major ATC centres, responsible for the Northern and Southern half of the country.

The goal of ATC work, and the tasks of air traffic controllers, is to maintain separation between aircraft in a way that is safe and allows for expeditious flow of air traffic. Air traffic controllers both direct the flow of traffic and provide in-flight information to assist aircrew in the operation of their aircraft. Work is divided between:

- *The Tower*, which provides airport control and surface movement control, and
- *The Area Approach Control Centre* (AACC), which provides 'Approach' control (responsible for aircraft approaching and departing the airport), 'Enroute' control (aircraft travelling to and from their destination), and 'Arrivals' control (preparing for landing).

In the AACC, controllers sit at a console providing them with information on their designated airspace of responsibility. The technology used by the controller at the console is known as TAAATS (The Australian Advanced Air Traffic Control System). It contains three main tools or instruments. First is the communications panel, where telephone and radiotelephone communications enable controllers to talk to pilots and controllers on other sectors. Second are the display mechanisms, which include air situation displays of aircraft tracks, representing electronic flight plans and trajectories. When radar is available, the radar displays, shows a trail of 'blips' or dots on the screen which represent a particular trajectory of a flight, with data alongside showing the flight number and flight level (altitude). In Australia, radar is available for air traffic on the Eastern seaboard only and around each major airport. When controllers separate aircraft in non-radar sectors, they use computer-generated displays anticipating where the aircraft is, based on information fed into the ATC system. Third are the map displays, which are used as a support for spatial reference. Although the controller is responsible for ensuring the safe conduct of the flight throughout that controller's airspace of responsibility, considerable collaboration occurs between the pilot and controller and with other controllers to ensure the traffic moves in a way that is orderly and expeditious.

Experience of ATC Work Practice

When the air traffic controllers interviewed described their work practice, it was found that they experienced this work along three dimensions that were present in the way the work was organized. The experience of ATC shaped – and was shaped by – these three dimensions. Work was experienced temporally, complexly and interdependently. All work can be mapped along these dimensions: that is, against the degree to which work must be undertaken in real-time or whether or not it can be started and stopped at will; whether problems to solve in work are simple or complex; and the degree to which work can be undertaken autonomously or is reliant on others. In ATC these dimensions were reported as being experienced in the following ways.

Temporal dimension

ATC work is structured so that the temporal nature of experience is emphasized. Temporally demanding work occurs in a dynamic 'real-time' environment, involving periods of intensity and requiring skills in concentration and immediacy. Because of the intensity and the immediacy, controllers reported that they often registered this dimension corporeally.

Complexity dimension

Second, ATC work is organized so that it is also experienced complexly. Complex work requires the coordination of multiple tasks that in turn require higher order thinking and – in combination with the temporality of work – an awareness and understanding is needed of the various permutations of problems and solutions that successful task completion may require within a limited period of time. Moreover, the knowledge base for complex and sometimes risky operations is inherently imperfect. The possible consequences of some of the actions involved in the work are sometimes quite cognitively demanding and this can be taxing.

Interdependent dimension

Finally, ATC work was reported as being experienced interdependently. When operating within complex environments, human operators cannot handle and hold in their minds all of the relevant knowledge, or consider all potential relevant trains of thought. They consequently have a reliance on each other and on artefacts such as tools and notation systems. That is, work is shared and is organized such that each individual is interdependent with others who form groups that in turn are interdependent with other groups.

The next section uses these dimensions as a framework to examine the experience of ATC work practice and its organization. Through this analysis, the attributes of work practice that lead to a greater likelihood of near-misses and mistakes are revealed. The resources employed to guard against these also provide insights into operations in High-3 work environments. The key dimensions, their attributes and their intersection are illustrated in Figure 8.1, which also includes examples of the resources used to mitigate risk.

The Temporal Nature of High-3 Work

In discussing the temporal coordination of work, Bardram (2000) notes that time in a workplace is one of the scarcest resources and its use needs to be prioritized

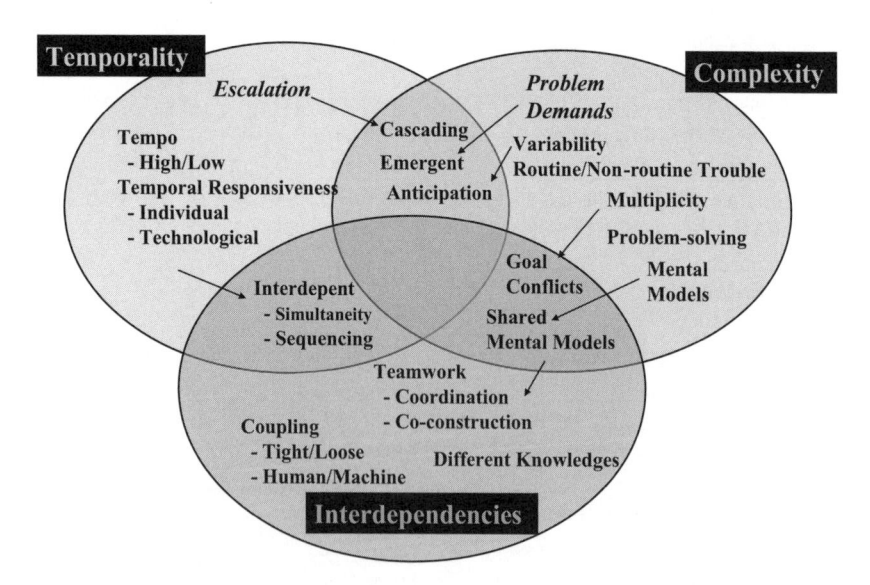

Figure 8.1 Intersection of temporal, complex and interdependent nature of High-3 work practices

and organized. Within High-3 work environments time is not just a scare resource, but also one where often the temporal flow of the work cannot be stopped. Once commenced, events must be managed as and when they occur. For air traffic controllers, this was the key element that distinguished their work from others. Three attributes relating to the temporal organization of work practice were identified: tempo, escalation and responsiveness (see Figure 8.1). A definition for each of these attributes is provided in Table 8.1.

Table 8.1   Dimensions of High-3 work, their elements and the resources employed to mediate risky work practice in air traffic control

Dimension	Definition	Resources
Temporal		
Tempo	The rhythm, pattern and speed at which operations are occurring (e.g., fast, slow)	Monitoring; body clock; technological alerts and timers
Escalation	The transition between changes in tempo	Self-awareness – third ear; teamwork; referential anchoring
Responsiveness	Synchronicity between actions and requirements needed in the activity system when the tempo escalates	Teamwork, referential anchoring
Complexity		
Problem demands; Emergent variability	Problem-based trajectories that emerge and permutate through time	Expectation of surprise; bricolage; mental models
Cascading	Multiple disturbances leading to anticipated and unanticipated consequences	Mindfulness; war stories
Multiplicity and goal conflicts	Multiple tasks to be undertaken and resolved that set up competing goals	Technological decision making tools; war stories; epistemic networks
Interdependence		
Coupling	The functional interconnections between system parts requiring coordination	Feeding-forward and feeding-back; TCAS; Shared air situation display; heedful inter-relating
Interdependent goal conflicts	Conflict arising from different intentions between operators and/ or between operators and systems	Epistemic networks; co-construction,
Simultaneity	The process where more than one activity must happen at the same time	Shared mental models; implicit coordination; shared air situation displays
Sequencing	When the output of one activity is required by the next operator	Shared mental models; implicit coordination; shared air situation displays

Tempo

The tempo of the work refers to the rhythm, pattern and speed in which operations are occurring. A fast tempo requires controllers to act quickly and live within the moment.

The pace and the intensity of the work are managed by controllers who claim that they develop good concentration and thus short-term memory skills so that they can attend to the particular problem at hand and then move on to the next. This immediacy also involves responding to the rhythm of the work and controllers believed that to be successful, they needed to develop good 'body clocks'. By developing the 'body clock' the controller uses the body as a resource to assist in knowing at what time actions should occur. Although there are alarms and alerts that the controller can manually fix and can be used to provide a monitoring of time, these in turn take time to set up and, as such, it is not always easy to rely on them. The instructors interviewed also talked about breaking or 'ruining' such technological supports so that the trainee would develop his or her own body clock.

The tempo of work practice can be either slow or fast and both have been implicated in making mistakes (Durso and Gronlund 1999). When work occurs at a fast pace the sheer volume and rapidity of work means there is the potential to be distracted and miss important things. Team members are sometimes called on to help out under these circumstances.

Likewise when the tempo of the work slows down the operator becomes more relaxed and therefore less vigilant. A link between this attribute and the cognitive complexity of work practice (see Figure 8.1) will be discussed in the next section. Under both these circumstances (low/high tempo), and the transition between them (Escalation), mistakes are more likely. This is because either the pace is too fast or because, having succeeded in managing an intense burst of activity, the operator's level of vigilance does not remain appropriately connected with the temporal requirements of the task at hand. This is illustrated in Figure 8.1 and will be discussed later when considering the interconnections between the temporal and complexity dimensions of work experience.

Monitoring the transition between a slow and a faster tempo is important and is characterized by the second attribute identified within the temporal nature of the work activity: Escalation.

Escalation

Escalation is the term used to describe the transition period between when the flow of the work moves from a low to a high tempo. An issue here is whether the operator's temporal completion of the tasks is sufficient to reflect this change when the tempo escalates. The operator needs to 'shift gears' to keep up with the work. Failing to do so means that mistakes are more likely. This tension in work practice (that is, failing to keep up with the escalation of the work tempo) has been described as when the operator 'falls behind the plane' (Woods and Cook 1999, 154). That is, because the work cannot be stopped, the system that the operator is working within continues on and the operator then needs to catch up. The allusion here is similar to that seen in early movies involving that newly introduced technology: the motor car. The driver would step out of the car and for one reason or another, the car would start moving, continuing on without the driver, who would be running behind and then alongside in an effort to regain control of the wheel. When an operator falls behind the plane (or the traffic), considerable effort needs to be exercised to catch back up again with the momentum of the system.

The quote 'falling behind the plane' illustrates both the time pressures involved and the short-term nature of task completion. The time pressures, intensity of work and short-term duration of the workflow mean that the controller must be aware of when to shift gears to manage an escalation and also when performance limitations are about to be reached, so that he or she can act before this point is reached, since the work can get too busy for one person to handle. Before this occurs, a decision must be taken to 'split' the airspace the controller is responsible for managing into smaller components that can then be managed by two people. Controllers described the need to recognize the signs that such an escalation was occurring. These included starting to perspire and stumbling over call signs as indicators that they were beginning to get overloaded. Thus the build up in the rate of workflow must also be monitored.

Sometimes controllers can end up being caught out by the level of escalation and they can then become too busy to conduct the processes needed to split the sector and thus to reduce their workload by reducing the size of the airspace. Under these circumstances in particular, teamwork becomes an important resource. This will be discussed in more detail when considering the interdependent dimension of work experience (see Figure 8.1). Of interest here, however, are the ways in which controllers use the activity of others to indicate escalation. Controllers connect their work to the work of others through developing what they called a 'third ear' (that is, being capable of listening out and picking up cues from the work of others that has implications for their work).

For example, the detected increase in workload of a controller on an adjoining airspace sector at a neighbouring console provides a common referent that can be used by controllers to monitor likely escalation for them since the work that is building up on an adjacent sector is likely to be coming their way. In ATC work the visibility of workload on other sectors enables controllers to 'referentially anchor' their work in relation to what can be anticipated within their sector. The term 'referential anchoring', developed by Resnick (1993, 10) is useful to characterize the way not only the resource of using the hearing detected through a 'third ear', but also the ways in which shared objects and displays are used to build up awareness about the traffic flow.

In this way the temporal experience of work intersects with the interdependent dimension of work practice (see Figure 8.1) as controllers use these shared spaces and shared objects to monitor the anticipated escalation in the flow of traffic. Because the information about the activity of work is available to others, team leaders and colleagues can also anticipate when a controller working at the console is likely to need help.

Temporal responsiveness

Temporal responsiveness refers to the synchronicity between operator actions and the requirements of others in the air traffic system. Temporal responsiveness is timely when things happen as and when air traffic controllers think that they should, with the controller ready to initiate actions in response to work demands. In this respect the idea supports the concept of joint cognitive systems (see, for example, Woods and Hollnagel 2006).

The responsiveness needed to manage an escalation in the tempo of work is sometimes difficult for trainees to learn. This is because the state of readiness needed requires a balance between vigilance on the one hand and relaxation on the other. This will be further discussed in the next subsection in relation to work complexity, in part because as trainees experience routines in the work, they develop better cognitive resources associated with anticipation.

Lags in temporal responsiveness can be both human as well as technological. Humans experience lags in temporal responsiveness when they fail to switch gears to keep pace with the work flow. A technological lag in temporal responsiveness occurs when, for example, a machine-initiated action does not occur in the time that the operators are expecting that it should occur. In the ATC system, for example, problems of such a lag have been evident in the introduction of a new form of communication called 'Controller-Pilot Data Link' communication (CPDLC). This communication system operates in a similar way to email, rather than voice communication. Typically controllers communicate by voice with pilots via radio. However, sometimes this can be quite difficult, especially when aircraft are in isolated areas and out of normal radio frequency range. The technological innovation of CPDLC enables text messages to be sent by either party if the surface and airborne systems are appropriately equipped.

However, some airspace incidents have occurred when the email-like message has been sent but not received and yet the controller is expecting the pilot to act (for example, to climb or descend at a certain point). In this case near-misses have occurred due to an unanticipated time lag created by a tool not providing the needed action in a timely manner. Accidents have been avoided through the use of another technological resource known as TCAS (The Collision Avoidance System) because the pilot did not act in the way the controller was anticipating that they should. Human and technological lags in responsiveness lead to problems of escalation in other parts of the activity system, as well as to increasing operator complexity.

The temporal dimension and its relationship to near-misses and mistakes

The temporal organization of work practice is implicated in increasing the possibility of near-misses and mistakes in a number of ways. The intensity of work created by the demands involved with processing and coordinating information within a limited time frame sets up a fertile ground for mistakes. Failing to monitor the level of escalation in tempo can also create increased pressure; especially if the time it takes to notice an escalation means that options that were available at a slower pace are no longer available. Human and technological lags in responsiveness also increase vulnerability in systems.

Controllers use a number of tools to help them manage the temporal demands of High-3 work. These include their own bodies, support from others, displays and resources integrated into the technologies-in-use. These demands and the resources available have to be integrated to align with another dimension of work experience in operation: the sheer complexity of the work.

The Complex Nature of High-3 Work

All work practice involves cognition, which includes problem-solving and decision-making. High-3 work frequently also involves higher-order thinking which has been defined by (Resnick 1993, 1) as thinking, which is:

- non-algorithmic (that is, the path of action is not fully specified in advance);
- complex (the total path is not mentally 'visible' from any single vantage point);
- effortful (often yielding multiple solutions, each with costs and benefits, rather than unique solutions);
- involves nuanced judgement and interpretation, requiring the application of multiple, sometimes conflicting, criteria.

However, High-3 work is not just demanding because it involves higher-order thinking. It is also demanding because of the serious consequences attached to it should it go wrong. Within High-3 work a mistake may result in a threat to the environment and to the health or lives of people either working in the organization or affected by it (Strater 2005). In the analysis four attributes were identified that need to be successfully managed when working in High-3 work environments. These are problem demands; emergent variability; cascading; and multiplicity and goal conflicts (see Table 8.1).

Problem Demands

When work involves higher-order thinking in a context of potential significant consequence the demands felt by the operator can be considerable. This is something air traffic controllers need to find and manage – the balance between attending to the consequences of the work and the detachment that can sometimes occur – as the following controller explains:

> The thing is, I think, you will find that there are very few air traffic controllers who actually sit back there and think *'This is an aeroplane full of three hundred people!'* They are just blips on the screen. That is why ... because if they *did* feel that it was three hundred people, they probably couldn't do the job very well. Some people have gone so far as to detach from what we are dealing with [and] they see it as a video game or a pinball game and they bunch them [aircraft] up close together and say *'this will be fine, this will be fine, this will be fine'* and [they are] cutting down the margins for error. That is one of the things we have to change. Okay, they *are* aeroplanes with people on board. [But] we don't want them to go overboard and say *'Oh Christ, there are several hundred people on this aeroplane!!'* At the same time we don't want them to say *'it's just a blip'*. It [successful performance] is somewhere in between. A happy medium.

To operate in a high-reliability environment where the consequences of action may put lives at risk, operators need to both be aware of the danger but not dwell on it. This requires a constant sense of awareness and vigilance on the part of the operator and a pre-occupation with the potential for failure (McCarthy, Wright and Cooke 2004; Rochlin 1999).

Emergent variability

Another feature of complex work in High-3 environments is its variability. Given the possible combinations of options and potential problems, a task or situation will rarely have exactly the same combinations of features a second time.

While controllers operating in this work environment must be vigilant to the possibility of variability they also experience 'routine trouble'. 'Routine trouble' is the term Suchman (1996, 37) gives to 'the kind of contingency to which the normal operations is perpetually subject ... it discloses the accomplished nature of that order ... [for which] a collaborative production of activities must be pulled off'. Non-routine trouble or tasks that require problem-solving are those that are unusual in a situation and therefore require the operator to be mindful that not all might be as it seems.

The concept of mindfulness has been discussed in a range of contexts (see, for example, Hollnagel and Woods 2006; Weick et al. 1999; Woods 2006) to describe the constant sense of awareness and vigilance required by operators working in High-3 environments: 'Mindfulness is less about decision making, a traditional focus of organization theory and accident prevention, and more about inquiry and interpretation grounded in capabilities for action'. Indeed, individual and collective mindfulness are necessary features to enhance in High-3 work environments because of their unanticipated variability and the linkage between this and the temporal flow of the work (see Figure 8.1). Nevertheless, problem-solving and decision-making are part and parcel of the work environment and also need to be addressed in any analysis of work practice.

In ATC work, decisions are made and remade as information becomes available. One resource controllers use is the development of individual and shared mental models about the problem situation. For Mathieu et al. (2000), a mental model is a mechanism whereby humans generate descriptions of a system's purpose and form as well as explanations of system functioning on observed systems states and predictions of future systems. Mathieu et al. (2000) contend that there can be multiple mental models coexisting among team members at any given point in time and that these would include models of job role, task and technology.

Mental models are typically the province of literature associated with individual cognition and situation awareness. According to Endsley (2000), an awareness of an air traffic situation is a subset of three elements: how well the problem is perceived, how well the problem is comprehended and how well the understanding of the problem is projected into the future. Hence in Endsley's definition the complexity of the problem and its solution are linked to the temporal domain. According to Endsley, errors occur when operators do not correctly perceive a situation and this may occur because either the data knowledge was unavailable or was difficult to detect or perceive. Failure to correctly perceive a situation can also occur because operators may not have noticed relevant emerging patterns in the temporal flow of events and have thus not given them due attention. Under these circumstances there has been a failure to comprehend the nature of the situation. According to Endsley this can be a result of using an incorrect mental model.

Although the concept of situation awareness is an important one, the conceptualization by Endsley has been criticized because the representation is always regarded as an external one (Hoc 2001). In the chapter by Marc and Rogalski (this volume) attention is given to how operators juggle complexity based on a concept of 'cognitive compromise'. That is, how operators juggle the external risks in the dynamic environment with their internal representation and understanding of the situation. Moreover, Marc and Rogalski go on to empirically demonstrate how this shared understanding is developed as a collective resource, in a way that is similar to what Mathieu et al. (2000) calls a shared mental model.

Individuals working interdependently can hold different types of knowledge and mental models about a situation that can be both a constraint and a resource. That is, if different levels of experience and knowledge about a situation are collectively brought to bear such that a coherent understanding of the problem and possible solutions are reached, then those different knowledges will be a resource. If, on the other hand, different knowledge and experiences are shared but fail to add to a coherent account or to possible solutions then these differences simply add to the vulnerability of the situation. All problems have trajectories that emerge and change through time and this can be exacerbated if things go wrong.

Cascading

The intersection between the temporal flow of work and its complexity occurs through emergent variability (see Figure 8.1). Duffy (1993) suggests that temporal patterns are not fixed in time in the sense that a given set of actions does not always last a certain number of minutes. In this sense, the notion of variability links with the temporal notion of escalation (see Figure 8.1). The demands of work complexity are also implicated in the temporal flow of escalation. When complexity is cascading (that is, when there are multiple disturbances leading to unanticipated consequences) these problem demands will be exacerbated by the temporal demands associated with escalation. In the following quote an Enroute controller is talking about a trainee who will make more work for herself if she does not manage the air traffic in a particular way, leading to emergent problems cascading and cutting down her opportunities to remain temporally responsive to the tempo of the work:

> We told her that it doesn't matter if that isn't exactly what you think is the perfect thing to do. If it is safe, be confident in it and do it, because if you hesitate, then things will get worse. You've got to be confident in your memory, what you remember, how you approach things. You've got to achieve the pilot's confidence straight away. He doesn't give you a second chance. If you do something and he feels ..., like ... he's ... putting his life in your hands, and if your confidence isn't there, or he doesn't feel it's there, then he's going to be nervous. And that's going to make things more difficult, because he's going to want to know more information, then you're going to get further behind, and it steam-rolls itself. So if you're not confident, those around you and those you speak to will hear it and you'll undermine yourself, and your confidence gets worse.

The important aspect of work practice illustrated here is that the controller's actions to some degree influence the number of disturbances and the time they will take to resolve them. In so doing, an emergent problem will become more demanding and then require more time to resolve. Thus problems can cascade together to create more unanticipated problems, as the problem trajectory takes a particular course.

To successfully deal with emergent variability requires both appropriate situation awareness about the problem, resilience and containment (Woods and Hollnagel 2006; Weick 2001). Operators 'deal with what is in front of them through operations that have an emergent quality similar to the activity of bricolage' (Weick 2001, 109). The French word bricolage (which has no precise equivalent in English) means to use whatever resources and repertoire are at hand to perform whatever task one faces. Invariably the resources are less well suited to the exact project than one would prefer but they are all there is.

Controllers use individual resources of constantly checking and scanning the situation to notice unanticipated disturbances and act on them before they cascade. This involves having both temporal and cognitive awareness of the situation. There is a pre-occupation with the potential for variability to lead to failure. This is an important resource in managing the complexity and temporal flow of the work. In High-3 environments, Rochlin (1999) called this 'working with a continuous expectation of surprise'.

Under certain circumstances (for example, variable weather patterns) controllers extend the amount of airspace separation required between aircraft, describing it as 'adding a little bit for mum'. What this means is that, to feel more comfortable, the controller has created a little bit of extra slack in the sequence, just as he or she would if a family relative were on board the aircraft.

Collective memory is also a resource that controllers draw on to access previous experiences of unanticipated disturbances created through cascading. In ATC, collective memory is transferred through the narrating of 'war stories'. War stories are the stories of a controller's experience when something dramatic happens, perhaps because of a controller's performance (or lack thereof), system deficiency or an unexpected event. War stories are passed informally between controllers and across ATC centres. War stories are used to help illustrate both good and bad actions, right and wrong ways of operating. War stories have a dramaturgical quality. They often set up an 'us-against-other' struggle or battle of some sort and, therefore, are well characterized as 'war' stories. The 'other' may be another controller, the technology, the environment, oneself or a combination of these elements. The following Enroute controller explains the importance of using war stories as a learning tool:

> Every one will tell you *'One day I had two here and they did this and that'*. That's a great way of gaining ... if you can't see it yourself, if you can hear someone else's war story, and when it so happens you get something that [is similar] you think *'I remember hearing about that. I'm not going to stuff up like that'*.

War stories act as a form of collective memory and become resources used to guide action and are particularly useful in assisting the controller to develop resources useful in situations of cascading. They are important in both providing learning opportunities from vicarious experience of the experience of cascading and the mistakes that can be made through task multiplicity and goal conflicts.

Task multiplicity and goal conflicts

It can be argued that multiplicity of goals and conflicts – within and between individuals – are always present in work situations. This can occur as an individual operator attempts to resolve his or her work problems alone or occurs in relation to working with others. The focus in this section will be on the former aspect, and the latter will be discussed further in terms of the interdependent dimension of High-3 work.

Whenever multiple tasks must be performed there is the potential for that multiplicity to result in conflicting goals. Indeed, working in ATC is regarded as the prioritization and management of competing demands. Those demands all impact on the task at hand and can be implicated in increasing the likelihood for near-misses and mistakes. These include: responding to the requests from different pilots, prioritizing a particular sequence of activity, deciding one solution to a problem that will impact on implications for later ones.

Technological tools provide the operator with information about possible courses of action available in their decision-making and collaborative activities in a process somewhat akin to what Flach (1999) calls 'feeding forward and feeding back'. What Flach draws attention to with this concept is that in dynamic complex systems adaptation and coordination of action depends on the interface where 'signals fed back map unambiguously the appropriate decisions and control actions' (Flach 1999, 121) for the future. When information is used to 'feed information forward', that information becomes a resource that can model the dynamics of a system and thus be used to anticipate possible mistakes. When controllers use technologies embedded in TAAATS, they have available to them tools that can be used to project a decision in time (for example, to be able to 'see' where and by when an aircraft will be if given a particular trajectory). This information can then be used to determine the impacts of a likely course of action on other aircraft currently in the system.

Goal conflicts can also be set up by the arrangement of the tasks, the application of rules and by the division of labour (see Chapter 2 for exploration of tensions between these elements in activity systems). In the following example, the controller is acting as an instructor to a trainee and finds it difficult to balance the job roles of ensuring ATC tasks are completed and giving the trainee the opportunity to do the job themselves and to learn:

> The hardest thing at first [when being an instructor] is to be able to concentrate on what is going on. You literally have to teach yourself to actually keep a full picture of what is happening. *'This has to be done now, he didn't do it. I have got to remember that he didn't do it'*. That is the first thing that you have got to teach yourself. That is very hard – being able to keep [a handle on] what is going on.

In this instance, there is a competing goal of performing as a licensed controller and performing as an instructor in an on-the-job training situation.

The complex dimension and its relationship to near-misses and mistakes

The problem demands and variability found in High-3 work creates opportunities for mistakes to occur. The enormity of the job can impact on how the operator confidently handles the work, making the job more difficult. However, if controllers become blasé about their job responsibilities this too leads to considerable risk. Unanticipated disturbances can emerge and cascade together to create even more complex circumstances. Situations can become unmanageable if not contained.

The management of complex problems that impact on one another in real time sets up a dynamic resulting in multiple competing goals. How those competing goals are managed sets up different possibilities for near-misses and mistakes. Sometimes the tensions are a result of competing roles played by operators in the activity system. For example, there is considerable pressure placed on an instructor who sits beside the trainee but at the same time remains responsible for traffic in that particular airspace. Multiple competing goals occur not just for individuals as they undertake their work, but especially as they engage in that work in relation to others.

The Interdependent Nature of High-3 Work

When work is complex and occurs within a real-time temporal domain, multiple actors are almost invariably involved. This is because the task is so complex that no one single individual is capable of undertaking the entire job (Lave 1996; Pea 1993) and in so doing a successful work outcome is contingent on other operators performing their work within another part of the system (Dietrich and Childress 2004). Three attributes (see Table 8.1) are important in considering the interdependent nature of High-3 work:

1. coupling;
2. interdependent goal conflicts;
3. simultaneity and sequencing.

Coupling

Coupling refers to the functional interconnections between parts. Interdependent actions are either loosely or tightly coupled (Woods and Cook 1999). In interdependent and complex systems, coupling of tasks can occur directly between operators or may be mediated through technological tools and artefacts. When High-3 environments are tightly coupled (Perrow 1999) there is a strong relationship of dependence between two activities and little room for uncoordinated action.

The degree to which coupling is implicated in increasing risky work practice is in part linked to the degree of visibility of the coupling process. For example, when controllers use the CPDLC system previously discussed to communicate with pilots, an active connection has to be established between that particular aircraft and the responsible controller. Early in the establishment of CPDLC some mistakes occurred where the aircraft was incorrectly coupled to another controller operating in another airspace. In one 'war story' told in the interviews, this resulted in a message request from a pilot to climb being received and approved by the wrong controller (a controller operating in a nearby airspace but not the controller responsible for the aircraft). Following this, two things happened. First, the controller receiving and approving the message was mildly perturbed to observe that the aircraft he expected to climb continued to maintain its flight level. However, this was nowhere near the anxiety felt by the controller nearby who had not sent a message to the aircraft and who observed, seemingly without reason, a dramatic change in one of the aircraft in his airspace, as it moved off track and commenced climbing. That disturbances like this can be managed without accident is testimony to the high-reliability processes used by operators and built into the work. Nevertheless, the example also illustrates the importance of the intersection between coupling, temporal lags and problem demands (see Figure 8.1).

The focus of any interdependent work is also based on teamwork or more broadly coordinated activities. Interdependence also intersects with complexity in terms of goal conflicts.

Interdependent goal conflicts

When individual operators are solving complex problems and managing multiple tasks they have to prioritize those tasks and their possible outcomes. Often this process involves comparing one potential outcome with another and then coordinating that outcome with others. Interdependence and complexity intersect (see Figure 8.1) when controllers are involved in managing multiple competing goals resulting in negotiated talk and trade-offs.

For example, the air traffic controller's primary interest is managing competing demands placed on him or her to optimize traffic flow through a particular air space. However, in the course of flight, the pilot of an aircraft wishes to optimize the fuel efficiency and performance profile of the aircraft. This means, for example, that when an air traffic controller manages the flow of air traffic and directs a pilot in a landing sequence that will delay the aircraft's planned arrival, there is likely to be tension sometimes resulting in contestation.

Sometimes there is an interdependent goal conflict that pilots and operators are not aware of, such as if aircraft continue on a trajectory where they are in danger of either a breakdown of airspace separation or, if unchecked, a collision. The TAAATS system has within it a tool to detect short-term conflict alerts (STCA) that will alert the controller to a potential risk. This is another example of Flach's notion of 'feeding forward'. In addition, some aircraft have a collision avoidance system (TCAS) built into them as a last resort if the shared mental models between pilots and controllers have become uncoupled. These tools are a resource that alerts the pilots of the impending danger such that they can take corrective action.

Two aspects that are important in coordination are the simultaneity and the sequencing of activity (Bardram 2000).

Simultaneity and sequencing

Interdependent *simultaneity* is a term Bardram (2000) uses to describe the process where more than one activity must happen at the same time between operators. Interdependent *sequencing* refers to when the output of one activity is required by the next (Bardram 2000). These activities clearly have implications for the timeliness of the temporal flow of the work discussed earlier.

Interdependent simultancity occurs in an ATC environment when two controllers share a console and work collectively on a particular airspace. For example, one controller might be managing the traffic and the other is listening and writing down the coordination. In Australian air traffic control, interdependent sequencing is more typical than interdependent simultaneity. The resources of most importance here include both teamwork and shared mental models. The following tower controller talks about the valuable roles his team members play in everyday work involving interdependent sequencing:

> They will relieve the person of all the little things. It might involve keeping an eye on two particular aircraft while that person is looking at another couple of aircraft. It might be writing a time on a strip which is not operationally critical, but it is part of the job we need to do for stats, and to relieve that person of that little bit extra that that person [inaudible] part of that job. Just to know that somebody else is doing it. It might be moving a strip, it might be a nudge on the shoulder at the right time [as a prompt] to say [something].

The coordination involved in interdependent sequencing becomes implicit (Entin and Serfaty 1999) when controllers share the same mental model of the traffic flow and its requirements such that no explicit communication is needed.

In technologies, such as TAAATS, controllers share the same air situation display. This enables them to see the aircraft travelling between their respective sectors. When controllers share a mutual understanding of the task environment, the task itself and interacting controller's tasks and abilities, they can anticipate the information required and this is a resource used when both controllers are busy as it means that controllers will offer each other the information required before it has been requested. One resource that enables the development of these shared mental models is adaptive team training (Entin and Serfaty 1999).

In a study into this form of coordinated cooperative activity, controllers have discussed how and when they use the organization of the console and its relationship to their airspace to coordinate their actions, as the following controller explains:

> It wasn't long before I realized you could pre-empt a whole lot of things, just simply by linking each button [on the console – to the airspace]. The realization

was, eventually, that each button reflected a particular block of airspace there. As soon as that light flashed, you mentally thought, '*North west*'. Or '*block southwest*'. And not only did you identify the block of airspace, but by looking at what you were holding there, you could predetermine, to an extent, what was going to come. So you're already turning things inside out. If the call was from [a particular airspace sector] as you're about to answer it, you're looking at your own outbound track to see what you've got. You're making and predetermining. At the same time, [another airspace sector] rings, but you're not holding anything from [that sector], so straight away, you take the call that you've got flight-strips for ... When you do eventually take [the other call], they might be asking you for a footy score or '*Have you seen Jack or Fred?*' or '*What's Bill doing these days?*'

Interdependence intersects with complexity under these circumstances, especially when unanticipated disturbances occur (see Figure 8.1). When disturbances occur and begin to cascade all resources are brought to bear on a situation. This might involve calling on team members and anyone else who might have a contribution to make to the situation. Typically this occurs in what Weick et al. (1999, 101) called 'informal epistemic networks'. Under these circumstances people self-organize into ad hoc networks to solve a problem. These networks provide operators with a rapid pooling of knowledge and, once the situation returns to normal, the epistemic networks dissolve.

For Engeström et al. (1999), this activity is called 'knotworking and co-configuration'. The notion is characterized by a 'pulsating movement of tying, untying and retying together otherwise separate threads of activity. The locus of initiative chances from moment to moment in a knotworking sequence' (p. 346). The idea of knotworking and co-configuration is similar to the notion Weick uses, except that people may not be physically together. For example, co-configuration might involve multiple actors physically distributed throughout the activity system interdependently working together for a brief period of time (for example, controllers operating in different centres, and aircrew). These actors all have subjective expertise in understanding the complexity of the situation. Adaptation and learning relies on the fast accomplishment of this intersubjectivity. It therefore involves shared and overlapping cognitive tools such as mental models and has a temporal dimension, as the problem is played out in time.

The interdependent dimension and its relationship to near-misses and mistakes

When work requires intricate coordination, possibilities for errors abound. Interdependent goal conflicts, coupling and interdependent sequencing and simultaneity require operators to draw on a range of resources to reduce the likelihood of error. These include technological resources such as STCA and TCAS, shared air situation displays and tools that enable the feeding-forward of information. Collective resources, such as teamwork, epistemic networks and co-configuration are also used.

The outcome of effectively managing the temporal, complex and interdependent dimensions of High-3 work practices has best been summed up by Bentley et al. (1992, 129) who concluded:

> Taken as a whole, the system is trustable and reliable... Yet if one looks to see what constitutes this reliability, it cannot be found in any single element of the system. It is certainly not found in the equipment... Nor is it to be found in the rules and procedures, which are a resource for safe operation but which can never cover every circumstance and condition. Nor is it to be found in the personnel who, though very highly skilled, motivated and dedicated, are as prone as people everywhere to human error. Rather, we believe it is to be found in the cooperative activities of controllers across the 'totality' of the system, and in particular in the way that it enforces the active engagement of controllers, chiefs and assistants with the material which they are using and with each other. This constitutes a continuing check on their own work and a crosscheck on that of others.

Conclusion

This chapter has developed an alternative framework for examining the High-3 work practices based on a close examination of work experience. Through a close examination of High-3 work practice in ATC, three interconnected dimensions of work experience have been identified. The intersection of these dimensions highlights the micro-processes of work and the ways in which they give rise to creating opportunities for mistakes and through a more nuanced understanding of High-3 work, the resources employed to mitigate these risks are also revealed.

It is important to recognize, however, the embeddedness of each of these dimensions within High-3 work and the role of the operator in creating resilience to the threat of mistakes under these circumstances. By drawing on this information, I hope researchers and scholars will be able to identify ways in which these resources can be strengthened, thus creating safe and more effective work environments.

This chapter has therefore proposed an alternative view of error and risk analysis by closely examining the ways in which work practices interact to create contexts that result in risky work and systems that are therefore vulnerable to error. It has highlighted the individual and collective technological and organizational resources that can be brought to bear to mediate those dynamics.

References

Bardram J.E. (2000), 'Temporal Coordination', *Computer Supported Cooperative Work* 9: 157–87.

Barley, S.R. and Kunda, G. (2001), 'Bringing Work Back In', *Organizational Science* 12(1): 76–95.
Bentley, R., Hughes, J.A., Randall, D., Rodden, T., Sawyer, P., Shapiro, D. and Somerville, I. (1992), *Ethnographically-informed Systems Design for Air Traffic Control*. Proceedings of CSCW '92, ACM Conference on Computer-Supported Cooperative Work (Toronto, Ontario: ACM Press).
Endsley, M.R. (2000), 'Direct Measurement of Situation Awareness: Validity and Use of SAGAT', in Endsley, M.R. and Garland, D.J. (eds.) (2000), *Situation Awareness Analysis and Measurement* (London: Lawrence Erlbaum Associates).
Engeström, Y., Engeström, R. and Vahaaho, T. (1999), 'When the Center Does Not Hold: the Importance of Knotworking', in Chaiklin, S., Hedegaard, M. and. Jensen, U.J. (eds.) (1999), *Activity Theory and Social Practice* (Aarhus: Aarhus University Press).
Flach, J.M. (1999), 'Beyond Error: The Language of Coordination and Stability', in Hancock, P.A. (ed.) (1999), *Human Performance and Ergonomics* (San Diego: Academic Press).
Hartel, C.E. and Hartel, G.F. (1995), *Controller Resource Management – What Can We Learn from Aircrews?* (U.S. Department of Transportation, Federal Aviation Administration).
Helmreich, R.L. and Foushee, H.C. (1993), 'Why Crew Resource Management? Empirical and Theoretical Bases of Human Factors Training in Aviation', in Weiner, E.L. et al. (eds.) (1993), *Cockpit Resource Management* (San Diego: Academic Press).
Hollnagel, E. (2006), 'Resilience: The Challenge of the Unstable', in Hollnagel, E. et al. (eds.) (2006), *Resilience Engineering: Concepts and Precepts* (Aldershot: Ashgate).
Hollnagel, E. and Woods, D. (2006), *Joint Cognitive Systems: Foundations of Cognitive Systems Engineering* (Boca Raton: Taylor & Francis).
Lave, J. (1996), 'The Practice of Learning', in Chaiklin, S. and Lave, J. (eds.) (1996), *Understanding Practice: Perspectives on Activity and Context* (New York: Cambridge University Press).
MacPhee, I. (1992), *Independent Review of the Civil Aviation Authority's Tender Evaluation Process for the Australian Advanced Air Traffic System*. Canberra: Australian Government Publishing Service.
Mathieu, J.E., Goodwin, C., Heffner, T., Salas E. and Cannon-Bowers, J. (2000), 'The Influence of Shared Mental Models on Team Process and Performance', *Journal of Applied Pyschology* 85(2): 273–83.
Murphy, M. (1980), 'Review of Aircraft Incidents', in Cooper, G. et al. (eds.) (1980), *Resource Management on the Flightdeck: Proceedings of a NASA/Industry Workshop* (Moffet Field, CA: NASA-AMES Research Center).
Owen, C. (1999), *Learning in the Workplace: The Case of Air Traffic Control*, unpublished PhD dissertation (Hobart: University of Tasmania).

Owen, C, (2001), 'The Role of Organisational Context in Mediating Workplace Learning and Performance', *Computers in Human Behaviour* 17(5/6): 597–614.

Pea, R.D. (1993), 'Practices of Distributed Intelligence and Designs for Education', in Salomon, G. (ed.) (1993), *Distributed Cognition: Psychological and Educational Processes* (Cambridge, MA: Cambridge Univeristy Press).

Perrow, C. (1999), *Normal Accidents: Living with High Risk Technologies* (New York: Basic Books).

Rochlin, G.I. (1999), 'Safe Operation as a Social Construct', *Ergonomics* 42(11): 1549–60.

Weick, K.E. (1987), 'Organizational Culture as a Source of High Reliability', *California Management Review* 29(2): 112–27.

Weick, K.E. (2001), *Making Sense of the Organization* (Oxford: Blackwell).

Weick, K., Sutcliffe, K. and Obstfeld, D. (eds.) (1999), *Organizing for High Reliability: Processes of Collective Mindfulness. Research in Organizational Behavior* (Connecticut: JAI Press).

Westrum, R. and Adamski, A.J. (1999), 'Organizational Factors Associated with Safety and Mission Success in Aviation Environments', in Garland, D.J. et al. (eds.) (1999), *Handbook of Aviation Human Factors* (London: Lawrence Erlbaum Associates).

Woods, D. (2006), 'Essential Characteristics of Resilience', in Hollnagel, E. et al. (eds.) (2006), *Resilience Engineering: Concepts and Precepts* (Aldershot: Ashgate).

Woods, D.D. and Cook, R.I. (1999), 'Perspectives on Human Error: Hindsights, Biases and Local Rationality', in Durso, F.T. (ed.) (1999), *Handbook of Applied Cognition* (Chichester, NY: John Wiley & Sons).

Woods, D. and Hollnagel, E. (2006), *Joint Cognitive Systems: Patterns in Cognitive Systems Engineering* (Boca Raton: Taylor & Francis).

Zuboff, S. (1988), *In the Age of the Smart Machine: The Future of Work and Power* (New York: Basic Books).

Chapter 9
Conclusion: Towards Developmental Work Within Complex and Fallible Systems

Christine Owen

The contributions in this volume have added to the understanding of the positive roles humans play in enhancing safety in complex and fallible systems. They have done so through providing a variety of analyses, adopting a human-centred focus in contrast to a techno-centred one (see Chapter 1).

The sites of these analyses have been undertaken in high-performance and high-reliability work contexts within systems of complexity. The workplaces featured here have been involved in aviation, maritime and rail transportation, chemical and oil manufacturing, emergency despatch centres and emergency medicine.

However, the findings and implications of these contributions are not just applicable to these work environments. The issues and challenges raised here will be important in the future because, as discussed in Chapter 1, the changing nature of work requires increasing proportions of the workforce to face the conditions found in the specific workplaces represented here. It is not just specialized work environments that are becoming increasingly inter-dependent, mediated by complex technologies, undergoing work intensification and facing demands from ever more unforgiving political environments. These characteristics are now the feature of many work environments. The challenge for researchers, practitioners and designers is how to support workers in these contexts.

The contributions in this volume aid our understanding of these work environments because they articulate the kinds of processes and practices that support workers to meet the goals of ensuring that system safety conditions are maintained and improved (Hollnagel and Woods 2006, 352). Attention to these goals is not new and research in this domain has been growing for over 25 years. What is new is how the contributions here add to this body of scholarship, taking it a step further, laying the groundwork for a new research agenda.

In 1984 Perrow criticized accident investigators for being too narrow and for failing to take account of economic and other realities of influencing the systems under investigation when making recommendations (see Chapter 2 for further discussion). This raised subsequent challenges, such as the difficulty of drawing broad conclusions on the basis of analyses of single accidents. These challenges led to the development of analyses based on different accident models (Rasmussen 1997; Rasmussen and Svedung 2000; Hollnagel 2002; 2004; 2006). The work of Hollnagel (2002), for example, was particularly important in providing important directions in critiquing the construct of human error, urging scholars to move beyond error and to give attention to normal everyday operations. More recently, the research focus has moved to human-centred technology design (Vicente 2004), to resilience engineering (Hollnagel, Woods and Leveson 2006) and to the importance of investigating how work actually happens rather than how it is imagined (Dekker 2006).

One of the key conclusions one can draw from these developments is that systems will always be flawed and in need of human intervention. Humans, too, are flawed and in need of support. However, is this all we can do? – ensure people and systems are safe from mistakes? Such a goal is, of course, laudable and critically important but is that all there is to it? Can we do more?

One of the implications from the contributions in this volume is that there is much more that, as researchers, designers, and practitioners, we can do. The challenge for the future is to develop strategies that not only help create safe environments and support operators working in fallible systems, but also to do so in ways that proactively contribute to the positive development of those people and their systems. Building work environments to enable growth and expansiveness in capacity and capability is not a new concept. What is surprising is the limited attention this goal has received in the literature.

Another of the the implications from the chapters contained here is that in safety-critical systems in particular, if we are to take this call seriously, then the ways in which workers are able to intervene, the resources they are able to draw on for support, and the subsequent judgements made of their practices – all require radical reassessment.

Through their analyses, the contributors in this volume aid in the development of this agenda by articulating the processes that support enabling developmental work.

Building a Research Agenda for Developmental Work

Developing a research agenda to support developmental work is important for two reasons. The first reason is a moral one: work is an essential part of life. It provides individuals with the opportunity to contribute to meaningful social relations and to the common good of society. Meaningful work facilitates the growth and development of the individual because 'workplace efficacy is linked to efficacy in all other areas of human activity' (Welton 1991, 10). For many people there is a close association between what they do everyday at work and the quality of their life in general. For example, authority structures in work organizations that lead individuals to undertake dull, repetitive work that blocks the imagination can lead to psychological conditions of helplessness and alienation (Welton 1991).

The second reason is an economic and political one: it is only through learning adaptation and change that organizations can adapt to changes in their environment and continue to develop and grow. This is particularly important in environments increasingly characterized by change and uncertainty. As organizations continue to become complex within ever more turbulent environments, learning and development in work is an important means to assist organizations to survive and flourish.

Drawing on approaches to work practice and analysis in particular (see, for example, Kornbluh and Greene 1989; Engeström 1999, 2001, 2005; Daniellou 2004, 2005; Nonaka and Takeuchi 1995), the goal of developmental work is to design, maintain and redesign work processes (for example, decision-making, technologies and organizational cultures) to enhance learning and to evaluate work relationships for their individual and collective learning and knowledge-creation potential.

The implications from the analyses presented here point towards some key aspects that are important in understanding what enables and constrains developmental work activity. For the purposes of this conclusion, the themes emerging in this book that may be employed according to this agenda are synthesized in Table 9.1 as pre-requisites, contexts and conceptual resources.

Key Pre-requisites for Enabling Developmental Work

One of the key starting points for enabling developmental work is being aware of the myriad of possibilities for framing up problems in the first place. The importance of naming up how problems are framed was explained by Wackers in the introduction to Part I. He pointed out that framing is a recursive process and as such the frame shapes what it is that we look for and what we see. Being conscious of the possibilities and constraints of the frame employed is important to ascertain what is being brought into focus and what is being left out.

A second starting point is an analysis of the degree to which problems can be made visible in the work environment. The possibilities and constraints of problem visibilization were discussed by Béguin, Owen and Wackers in Chapter 1.

Table 9.1 Theoretical resources for developmental work activity

Pre-requisites	Contextual conditions	Conceptual resources
Framing Problem visibilization • Reflexivity • Learning Aligned research methods	Tension, stress and contradiction Constant change	Migration and drift Developmental trajectories Boundaries Analysis of interdependencies Building plasticity

As discussed, if problems are unable to be visibilized, learning and change cannot occur. Problem visibilization will be difficult in environments that are hostile, have limited reflexivity and cultures that do not support speaking up. They will also be constrained by poor data collected on work practice. Pre-requisites are environments where problem visibilization is enabled and where reflectivity is encouraged. These two pre-requisites (problem visibilization and reflectivity) are needed so learning and change can occur.

A third starting point is to choose research methods sympathetic to a human-centred focus. As Béguin pointed out in his introduction to Part III, there are two possible paths for knowledge development: a logical-reductionist mode and the narrative mode emphasising meaning and interpretation. Though by no means representing the only methodologies that are available, the chapters here present rich narratives to uncover meaning workers give to their everyday practice. In addition to the methods employed here are intervention-based methods associated with encouraging workers to interrogate their own conditions and practices (see, for example, Daniellou 2005; Engeström 2005).

Contextual Conditions of Contempory Work

There are a number of themes suggested in the chapters that point to how we may articulate a developmental work agenda. Two of these provide a background for understanding the context of the work environment and the remaining are used as conceptual resources for interrogating the work.

Tensions and contradictions

One of the underpinning arguments present in many of the contributions is that of acknowledging that in work environments there are always tensions. Indeed, many of the contributors here have a connection with cultural-historical activity theory (see Chapters 2, 6 and 7) where tension and contradiction is identified as a key driving force for change. Many of the contributions in this volume call for a recognition of the need to live with stress in systems. That is, to acknowledge that tensions will always be part and parcel of working life.

This is in contrast with much of the human factors literature, where tension and stress in systems is perceived as something that either must be eradicated or as a particular state of which systems drift in and out. One of the implications of the analyses discussed here is to suggest that tensions, contradictions and stresses are always embedded in complex systems. Norris and Nuutinen in Chapter 2 make a point of systematically analyzing the systemic tensions and their influences on human action. Rosness (Chapter 3) examines the systemic tensions embedded in the evolution of organizational development and procedures. For Messman, in Chapter 5, there is always the tension of adjusting to work as it happens. Her argument is similar to that developed by Marc and Rogalski (Chapter 6), who analyze the trade-offs operators make between ensuring safety and efficiency. Finally, Owen (Chapter 8) conceptualizes how, in the context of work activity, problem demands need to be negotiated within temporal and interdependent constraints.

Therefore, we need to find ways to live with and to articulate what those stresses, tensions and contradictions are so that problems may be visibilized. This also means we need to reframe how we think about change.

Change

The conclusions drawn in many of the chapters suggest a constancy of change. That is, we will never reach an ideal state, since one does not exist. The implication of this is that we need to normalize how we work within and manage constant states of transition and how we live with and grow within contexts of constant transition and change.

Tools for Diagnosing Developmental Work Possibilities

A number of contributions provide an analysis of change to reveal how history is embedded in present action. Wackers, in Chapter 4, shows how trends in global capital affect pricing of oil commodities, which in turn influences workers involved in its extraction. Rosness, in Chapter 3, provides an analysis of how human action is compromised in states of transition and impeded by organizational history. Mesman's argument, in Chapter 5, is that staff working in intensive care units have a continuous struggle to manage unpredictable changes that result in sometimes erratic flows of action.

One of the analytical tools employed by contributors includes the notion of migration or drift. The three chapters in Part I all draw on conceptualizations of drift or migration to understand how safety-critical systems become vulnerable to accidents. All three draw in particular on the work of Rasmussen's 'drift to danger' model (see, for example, Rasmussen and Svedung 2000).

A second conceptual resource employed in the analyses includes that of boundaries. In the case of Norris and Nuutinen, in Chapter 2 and Rosness, in Chapter 3, boundaries are important because they constitute the limits of safety. The question is how are these boundaries visibilized? And in particular, how are they visibilized in contexts of constant change? Norris and Nuutinen draw on Reiman and Oedewald's (2007) metaphor of the 'safety envelope' and also on Dekker's concern to keep the boundaries of safety visible. For Dekker (2006, 75–92), markers of organizational behaviour that support keeping boundaries visible are active discussion about risks in the organization, and evaluating how an organization responds to failure.

Keeping boundaries visible is of concern to Rosness in Chapter 3. He asks whether boundaries of safe performance are easily visible to the actors. 'What will happen when an activity approaches or crosses the boundary? Will the actors receive an insistent warning from the system and have the opportunity to reverse their actions?' Since many dangerous situations do not lead to disaster, there a risk that they will adapt to warnings over time.

In addition, the contributions in this volume analytically utilize the concept of the boundary in different ways. How boundaries shift, for example, is of interest to Norris and Nuutinen (Chapter 2) who use it as a theoretical tool through which to identify tensions as driving forces for change. For Norris and Nuutinen, analyzing the ways in which the boundary of safe action shifts due to systemic tensions is of particular importance in their model. In their model, Norris and Nuutinen drew connections between the efficient system-related reasons that were influencing how people acted. In doing so they used the notion of practice that enabled the connection to be made between human action and broader systemic constraints and contents.

Finally, boundary blurring was raised as a challenge by Béguin in his introduction to Part III. He suggests it is not only boundaries that should be visibilized, but that in some contexts safe action is enhanced if indeed boundaries are blurred. For him safe situations depend on the removal, or at least the blurring, of boundaries between different occupational groups. A constraint in enabling this to happen is the strong division of labour that can be found in many workplaces. A strong division of labour can be a barrier to shared understanding of a situation because different perspectives are based on different contextual social positions, knowledges and interests. Béguin (Part III introduction and Chapter 7) uses the concept of 'world' to grasp this point, using ideas that come from Cassirer's 'science of culture'. He maintains that in his case, two different social positions – those of the user and those of the designer – constitute two different 'worlds', which never meet. This is a theme also addressed in Part I (in terms of collaboration across boundaries) and in Part II (in terms of the role of common artefacts). The challenge then under these circumstances is how we might bring forward in the design of complex and fallible systems shared understanding so that learning can occur and that weaknesses can be both identified and mitigated. A third conceptual resource used in the contributions is that of trajectories.

Trajectories of work development

A number of contributions analyze change by tracing systemic trajectories over extended periods of time (see, for example, Chapters 2, 3 and 4). The question these chapters raise for practitioners is what are the developmental trajectories of the work activity under study? Understanding history and how layers of history are embedded in the present allow us to chart a path into our future, by recognizing the horizons and constraints of possibility.

In addition some of the contributions also employ the concept of trajectories-in-action (see, for example, Chapters 5 and 8). Mesman, in Chapter 5, reminds us that the care trajectory of critically-ill newborn babies is always uncertain and to some degree will always have features that are unique. In her work there is an explicit connection between trajectories and tensions: between the need for intervention and making adjustments on the one hand and the associated risks and uncertainties on the other. This trajectory of action and its inherent tensions with inaction is close to that analyzed by Marc and Rogalski (Chapter 6) in their conceptualization of cognitive compromise and ecological safety. Marc and Rogalski's development of the model of ecological safety highlights the relationships between safety, risk management and error management. It shows the ways in which operators engage in strategies of cognitive compromise, weighing up internal risks and external risks. They also demonstrate the way in which operators draw on the collective as a resource.

Connected to understanding trajectories in action is analysis of the temporal dynamics of work in complex systems. A number of contributions analyze the function of time in action. The taxonomy of work practice modelled by Owen (see Chapter 8), for example, is more than simply challenges of coordination. The contributions here analyze the way timeframes can be disrupted, shift out of phase (see Chapter 8) or implode, and how 'temporal niches' (to create space for decision-making) can be shaped (see Chapters 5 and 6).

Analyzing how work practices develop

The challenge of how change occurs is something not frequently addressed in the literature. Both Norris and Nuutinen (Chapter 2) and Owen (Chapter 8) employ the notion of practices to analyze change in work contexts.

Norris and Nuutinen, for example, employ the habit-based concept of practice (see, for example, Norros 2004), to define and connect practices to systemic modelling of the domain concept. In the case of Norris and Nuutinen, practices are built between systems and human action that develop through habit. Adopting the notion of practice in this way allows an empirical analysis of activity in specific situations so that performance can be evaluated with relation to broader systemic constraints.

Norris and Nuutinen also discuss the brittleness of practices – and formulated predictions concerning the possible brittleness of practices when the conditions of maritime piloting become more demanding. In this respect the theoretical resources mentioned previously come together, since it is important to be able to visualize the trajectories that lead to boundary shifting. The challenge for the future is how conditions of drift might become visibilized, particularly those conditions that may change practices so that they become brittle and thus lead to increased vulnerability in systems?

A key challenge for the future is to understand what hinders the development of practices. As Rosness (Chapter 3) describes, we need to better understand how practices degrade in systems over time and how, as Norris and Nuutinen suggest, practices become brittle. For Owen (Chapter 8) practices develop over time through collective memory. What is also important in this respect is to acknowledge that the responsibility for enhancing practice is held collectively as well as individually, which is different from the usual approach taken in the literature. The degree to which people can learn from, for example war stories in air traffic control culture (see Chapter 8), depends on access to those with the experience-based narratives. Access to war stories is access to previous experiences of unanticipated disturbances.

Mesman (Chapter 5) also shows how collective memory is employed to turn a near-miss into new practices for the next newborn. In the contexts discussed in these chapters, practices develop in the space between systemic structures and human improvisation. They endure and become robust through learning held in collective memory.

Analyzing interdependencies and the role of the collective

The ways in which individuals are able to support each other and what enables and constrains their collective work are critically important. A number of chapters illustrate socio-cultural approaches to understanding the ways in which collective performance is embedded in the context of the work environment and not separate from it. Marc and Rogalski (Chapter 6), for example, show how observed interactions are oriented toward a shared situation assessment, enabling recovery of errors through actors mutually cross-checking each other's information assessment and operations. In their chapter Marc and Rogalski illustrate how operators simultaneously function at three levels – at an individual level: one of producing and managing one's own errors; at the social or team level, where the individual operator manages the errors of others and at an organizational level, where the team is operating as a collective unit or described as a 'Virtual Operator' (to distinguish from the individual operating autonomously or as part of a team). Being able to analyze activity at the level of the collective is a significant development.

Similarly, Owen illustrates the ways in which interdependent action is mediated by a range of other elements embedded in the work context. These include temporal demands as well as task demands, each having an interactive effect on coordination and work practice.

Building plasticity to support developmental work

A common theme raised by many of the contributors in this volume is the challenge between what is fixed in procedure and what is able to be improvised in action. Marc and Rogalski (Chapter 6) analyzed the role played by observers in facing collective work with regards to operational memory and shared situation awareness. When participants observed possible departures from prescribed procedures the status of those departures was questioned. Are they and should they really be considered as errors?

For Mesman (Chapter 5), there is a tension between what cannot be formalized as knowledge and codified as protocol. In understanding the complexity of a trajectory of care needed for a critically-ill newborn, Mesman argues that complex systems like the NICU must 'eschew rigour, sternness and harshness. Instead, protocols have to be negotiable, facts have to be malleable, and time has to be makeable'. For Béguin (Chapter 7) the critical issue is how designers can allow for the development of plasticity in work to enable, as he suggests, users to finish the design in the context of their own work practice.

Where To From Here?

The chapters in this volume provide a rich theoretical and empirical understanding of human work in complex and fallible systems and of the positive roles humans play. Through their discussions we can begin to articulate and develop a new research agenda – one that is more than just ensuring humans do not make mistakes, or identifying strategies needed to make systems safe.

To enhance the positive role humans can play in ensuring safety in fallible systems it is important to identify the features of developmental work. It is clear, however, that there are many impediments to enabling workplaces to become developmental environments, both from a human-action and a system-safety point of view. These constraints need to be confronted.

The reader's challenge, using the conceptual tools presented here, is to consider what the outcomes of an evaluation of work within their own contexts would yield in terms of insights and understandings about the potential developmental trajectories for future action. It is time to begin.

References

Daniellou, F. (2004), 'Alain Wisner: Learning from the Workers around the World', *Ergonomia* 26(3): 197–200.
Daniellou, F. (2005), 'The French-speaking Ergonomists' Approach to Work Activity: Cross-influences of Field Intervention and Conceptual Models', *Theoretical Issues in Ergonomic Science* 6(5): 409–427.
Dekker, S. (2006), 'Resilience Engineering: Chronicling the Emergence of Confused Consensus', in Hollnagel, E. et al. (eds.) (2006), *Resilience Engineering: Concepts and Precepts* (Aldershot: Ashgate).
Engeström, Y. (1999), 'Expansive Visibilization of Work: An Activity Theoretical Perspective', *Computer Supported Cooperative Work* 8(3): 63–93.
Engeström, E. (2001), 'Expansive Learning at Work: Toward an Activity Theoretical Reconceptualization', *Journal of Education and Work* 14(1): 133–56.
Engeström, E. (2005), *Developmental Work Research: Expanding Activity Theory in Practice* (Berlin: Lehmanns Media).

Hollnagel, E. (2002), 'Understanding Accidents: From Root Causes to Performance Variability', *IEEE 7th Human Factors Meeting* (Scottsdale, Arizona).
Hollnagel, E. (2004), *Barriers and Accident Prevention* (Aldershot: Ashgate).
Hollnagel, E. (2006), 'Resilience: The Challenge of the Unstable', in Hollnagel, E. et al. (eds.) (2006), *Resilience Engineering: Concepts and Percepts* (Aldershot: Ashgate).
Hollnagel, E. and Woods, D. (2006), *Joint Cognitive Systems: Foundations of Cognitive Systems Engineering* (Boca Raton: Taylor & Francis).
Hollnagel, E., Woods, D. and Leveson, N. (eds.) (2006), *Resilience Engineering: Concepts and Precepts* (Aldershot: Ashgate).
Kornbluh, H. and Greene, R. (1989), 'Learning, Empowerment and Participative Work Processes: The Educative Work Environment', in Leyman, H. and Kornbluh, K. (eds.) (1989), *Socialization and Learning at Work: A New Approach to the Learning Process in the Workplace and Society* (Brookfield, VT: Gower).
Nonaka, I. and Takeuchi, H. (1995), *The Knowledge-creating Company: How Japanese Companies Create the Dynamics of Innovation* (New York: Oxford University Press).
Norros, L. (2004), *Acting Under Uncertainty: The Core-Task Analysis in Ecological Study of Work* (Espoo: VTT Publications No. 546). <http://www.vtt.fi/inf/pdf/publications/2004/P546.pdf>
Perrow, C. (1984/1999), *Normal Accidents: Living with High-Risk Technologies. With a New Afterword and a Postscript on the Y2K Problem* (Princeton, NJ: Princeton University Press).
Rasmussen, J. (1997), 'Risk Management in a Dynamic Society: A Modelling Problem', *Safety Science* 27(2/3): 183–213.
Rasmussen, J. and Svedung, I. (2000), *Proactive Risk Management in a Dynamic Society* (Karlstad, Sweden: Swedish Rescue Services Agency).
Reiman, T. and Oedewald, P. (2007), 'Assessment of Complex Sociotechnical Systems: Theoretical Issues Concerning the Use of Organizational Culture and Organizational Core Task Concepts', *Safety Science* 45(7): 745–68.
Vicente, K. (2004), *The Human Factor* (New York: Routledge).
Welton, M. (1991), *Toward Development Work: The Workplace as a Learning Environment* (Geelong: Deakin University Press).

Index

Bold page numbers indicate illustrations, *italic* numbers indicate tables.

A

Accident Investigation Board Finland (AIB) 18–19
accidents
    as always negative occurrences 17
    dissatisfaction with study of 4
    models 19–21
action
    and decision-making, decoupling of 71–2
    temporality and trajectories in negotiating 102
activity system perspective on piloting 40, **41**, 42–3
adaptive processes, fast and slow levels 85–6
adaptiveness 25
adaptivity 14
air traffic control (ATC)
    bricolage 186–7
    cascading 186–7, 192
    co-configuration 192
    cognitive compromise 185
    collective memories 187
    Collision Avoidance System, The (TCAS) 182
    complexity dimension 177, **178**, *179*, 184–9, 192
    Controller-Pilot Data Link communication (CPDLC) 182
    coupling 189–90
    decision-making and mental models 185
    emergent variability 184–6
    escalation of work 180–2
    falling behind the plane concept 181
    goal conflicts 187–8, 190
    interdependent dimension 178, **178**, *179*, 189–93

    knotworking 192
    mindfulness 184–5
    near-misses 189, 192–3
    organization of 176–7
    problem demands 184–9
    referential anchoring 181
    sequencing 191–2
    simultaneity 191–2
    task multiplicity 187–8
    technology and time lags 182
    tempo of work 180
    temporal dimension 177, **178**, 178–83, *179*
    temporal responsiveness 182
    war stories 187
Amalberti, R. 129, 133
Antolin-Glenn, P. 134
Åsta accident on Rørøs line, Norway
    acoustic collision alarms 65, 71
    Automatic Train Control on Rørøs line 63–6
    boundaries of safe performance 68–9
    breakdown in information flow 70–1
    causal and contributory factors 66, **67**
    conflicting demands on railway system 68–9
    decoupling of decision-making 69–70
    departure procedure 64–5
    event sequence 53, 56–8, **57**
    evolution of vulnerability in system 15
    evolution of vulnerability on the Rørøs line 62–6, 76
    human contribution to risk control 76–8
    learning from 74–8
    missing defences 58–9
    normalization of deviance 73
    problem-solving and structural change 71
    reframing the causal analysis 66–74
    second representation of 73–4, **75**
    status of mobile phones 65–6

structural change and problem-solving 71
traffic control on Rørøs line 59, *60–1*, 61–2
aviation industry
  historical context of risk-taking in 174
  need for requisite variety 175
  studies of 175–6
  see also air traffic control (ATC)

B
Bardram, J.E. 178, 191
behaviour assessment from practice viewpoint 33–4
Berg, M. 114
Bødker, S. 158
boundaries of safe performance 7, 20–12, 68–9
breakdown in information flow 70–1
bricolage 186–7
Bruner, J. 151
Brunsson, N. 71–2

C
Callon, M. 12, 84
Carroll, J.M. 157
cascading 186–7, 192
Cassirer, E. 159
catachreses 155–6
Challenger space shuttle explosion 73
change, constancy of 201
Clot, Y. 46
co-configuration 192
cognitive compromise 129, 132–4, 185
collective activity
  error and safety management in 134, 142–5
  in an emergency medical centre 135–7
  individual error management in 137–9
collective memories 187
Collision Avoidance System, The (TCAS) 182
common worlds
  in the design process 158–60
  through mutual learning 160–2
complexity and coupling 55
complexity dimension of air traffic control (ATC) 177, **178**, *179*, 184–9, 192

compromises between external and internal risks 133
consistency 133
constancy of change, 201
Controller-Pilot Data Link communication 182
Cook, R.I. 23
cooperative design approach 158
Core Task Analysis (CTA) 26–7, 29, *30–1*, 31–3, 44–5
coupling 54–5, 189–90

D
daily work, risky work studies on 18
decision-making
  and action, decoupling of 71–2
  and mental models 185
  and protocols in neonatal intensive care units 114–16
  decoupling of 69–70
  incremental, and the need for watchdogs 72–3
decoupling 14
  of decision-making 69–70
  of decision-making and action 71–2
Dekker, S. 22
departure procedure on Norwegian railways 64–5
design, failure of 92
design process
  as mutual learning process 157–8
  catachreses 155–6
  common worlds 158–60
  common worlds through mutual learning 160–2
  cooperative design approach 158
  instrument-mediated activity approach 153–7
  instrumental genesis 156–7, 164–5
  levels of learning 168
  mutual learning and interdependence 157
  redefinition of project 165–6
  safety instrument system 162–7, *164*
  simulation of activity 166–7
developmental work
  analysis of interdependencies as theme 204

collective work as theme 204
constancy of change as theme 201
development of practices 203–4
plasticity 204–5
pre-requisites for enabling *199*
research agenda 198–9
stress in systems as theme 200–1
tools for diagnosing possibilities for 201–5
trajectories of work development as theme 202–3
deviance, normalization of 73
distributed decision-making 69
Doireau, P. 136
double-binds 5
double-checking by humans 62n1
drift
    as analytical tool 14, 201
    in sociotechnical systems 54–5
    to danger model 22

E
ecological safety approach 129, 132–3
Efficiency-Thoroughness Trade-Off 54
emergency medical centre
    case management 135
    cognitive compromise 132–4
    collective activity in 135–7
    consistency 133
    ecological safety approach 132–3
    error allocation 141–2
    error and safety management in 142–5
    external and internal risks 131–3
    field observations simulation 130
    individual error management in collective activity 137–9
    operational memory 139–40
    safety management 140–1
    shared mental models 140
    situation mastery 141–2
    sufficiency 133
Endsley, M.R. 140, 185
energy
    distribution of responsibility in industry 89–90
    financial markets as influence on industry 84–5
    humanity's hunger for 83–4
    licencing of oil companies 87–8
    long-distance control of industry 88
    Petroleum Act (Norway) 89–90
    slow levels of adaptive processes 86–7
    stacked rules and regulations of industry 88–9
    see also offshore energy production
Engeström, Y. 192
envelope of safe action 22–3
environment-human system 24–5
epidemiological models 20–1
error(s)
    dissatisfaction with study of 4
    in collective activity 134, 142–5
    individual management in collective activity 137–9
    persistence of as focus 2–3
    points of view of 130–1
escalation of work in ATC 180–2
evidence-based medicine 109n2
exnovation 125
experience
    developmental history of 14
    distribution of in neonatal intensive care units 121–5
expert identity construct 32–3
external risks 131–3, 135–7

F
falling behind the plane concept 181
Faverge, J.-M. 149
Finland. *see* piloting accidents in Finnish fairways
Flach, J.M. 188
framing 12–13, 84, 199
friction in systems. *see* systemic stress
functional modelling 22

G
gas production. *see* energy
generalization from accident analysis 45–6
goal conflicts 187–8, 190
Grønbæk, K. 158
Gunderson, L.H. 87

H
habit 24, 44

high-3 work
  complex nature of 183
  definition 173
  higher-order thinking 183
  vulnerability to risk 173–4
  see also air traffic control (ATC); aviation industry
high reliability organization (HRO) school 22
higher-order thinking 183
hindsight, paradoxes of 78
Hoc, J.-M. 129
Holling, C.S. 87
Hollnagel, E. 20, 22, 54, 197–8
human-centred focus on work 4
human-environment system 24–5
human error, move away from 13
humans
  contribution to risk control by 76–8
  positive role in fallible systems 6–7
Hutchins, Edwin 36–7

I
incremental decision-making and the need for watchdogs 72–3
information flow, breakdown in 70–1
instrument-mediated activity approach 153–7
instrumental genesis 156–7, 164–5
interdependencies, analysis of 204
interdependent dimension of air traffic control (ATC) 178, **178,** *179,* 189–93
internal risks 131–3, 135–7
investigations, systematic working as essential 19

J
Jentsch, F. 134
joint cognitive system 25

K
Kelmola, U.-M. 23
knotworking 192
knowledge and experience, distribution of in neonatal intensive care units 121–5
knowledge development 200

L
language as a common world 159
Law, J. 89, 93, 113
learning
  from accidents, difficulties with 17
  levels of in design process 168
  mutual, and interdependence 157
  neonatal intensive care units 120–1
  see also piloting accidents in Finnish fairways
Leveson, Nancy 21, 22
licencing of oil companies 87–8
Lindblom, C.E. 72
loose coupling 55

M
man-made disasters, theory of 70–1
Marc, J. 185, 204
marine transport systems as error-inducing 18
Marion, R. 87
Mathieu, J.E. 185
McCarthy, J. 29
medical emergency centre. *see* emergency medical centre
memory
  collective 187
  operational 139–40
mental models and decision-making 185
Mesman, J. 205
migration of boundaries 68–9, 201
mindfulness 184–5
mobile phones, status of in Norway railways 65–6
multiplicity of tasks 187–8
mutual learning
  and interdependence 157
  common worlds through 160–2

N
navigation, traditional/technologically mediated 37–8, **39, 40**
near-misses
  and complexity 189
  and interdependence 192–3
  definition 174
Nemeth, C. 23

neonatal intensive care units
  change and uncertainty in 105–6, 125
  decision-making and protocols 114–16
  deviations from protocols 119
  distribution of knowledge and experience 121–5
  exnovation 125
  focus on practice in research 107
  frameworks for data comparison 118, 119
  information and data in 111–14
  interaction between protocols and expertise 125
  interpretation and use of data 121–5
  learning from errors 120–1
  protocols in 106, 109–10
  temporal niches 115–16
  transition moments 116–18
Normal Accidents (Perrow) 45
normalization of deviance 73
Norros, L. 23, 24, 203
Norway
  acoustic collision alarms at Rail Traffic Control Centres 65, 71
  breakdown in information flow in railway system 70–1
  conflicting demands on railway operations 68–9
  decoupling of decision-making and action 71–2
  decoupling of decison-making in railway system 69–70
  departure procedures on railways 62–3
  railway organisation in 62–3
  status of mobile phones on the railway 65–6
  structural change and problem-solving in the railway system 71
  see also Åsta accident on Rørøs line
Nuutinen, M. 203

O
Oedowald, P. 22–3
offshore energy production
  distribution of responsibility in industry 89–90
  financial markets as influence on industry 84–5
  impact of oil prices dropping 90–1
  licencing of oil companies 87–8
  long-distance control 88
  ordering and coordinating work processes 92–5
  Petroleum Act (Norway) 89–90
  recursive process as mode of ordering 93–4
  regularity gradient 90–2
  Sleipner A incident 81
  slow levels of adaptive processes 86–7
  Snorre A incident 82, 95–6
  stacked rules and regulations 88–9
  Super Puma helicopter incident 81
  see also energy
oil production. *see* energy; offshore energy production
operational memory 139–40
organizational behaviour and resilience engineering approach 22

P
panarchy 86
patterns 25
Peirce, C.S. 24
performance closure 82
Perrow, Charles 4, 13, 18, 45, 54, 197
Petroleum Act (Norway) 89–90
piloting accidents in Finnish fairways
  activity system perspective 40, **41**, 42–3
  aims of investigation 27
  behaviour assessment from practice viewpoint 34–8, *35*, **37**, **39, 40**
  core-task modelling of pilotage 29, *30–1*, 31–3
  decision to investigate 27
  description of facts of accidents 29
  expert identity construct 32–3
  features of good piloting 32–3
  historical origin of difficulties 42–3
  path of accident investigation **28**
  reports from investigation 27
  resources assessment 33–4
  results of phases **28**
  scarcity of investigation material 46
  situational demands assessment 33–4
  tensions identified 42

traditional/technology-mediated
    navigation 37–8, **39, 40**
plasticity in systems 101–2, 154, 204–5
power relations 6
practical drift 55
practices
    concept of 23–5
    development of 203–4
    increasing resilience of normal
        44–5
    pilotage assessment 34–8, *35,* **37**
    rule-based notion of 92–3
Prigogine, I. 83
problem demands in air traffic control
    (ATC) 184–9
problem-solving and structural change
    71
problem visibilization 6, 200

R

railways
    acoustic collision alarms at Norwegian
        Rail Traffic Control Centres 65, 71
    conflicting demands on Norwegian
        operations 68–9
    decoupling of decision-making and
        action in Norway 71–2
    departure procedure on Norwegian
        railways 64–5
    organization of in Norway 62–3
    status of mobile phones in Norway
        65–6
    structural change and problem-solving
        in the Norwegian system 71
    see also Åsta accident on Rørøs line
Rasmussen, Jens 4, 7, 13, 20, 22, 45, 54,
    91, 93, 136, 150, 153
Reason, J. 90, 91–2
recursive process as mode of ordering
    93–4
redundancy 102
referential anchoring 181
regularity gradient 87, 90–2
regulatory institutions 72–3
Reiman, T. 22–3
requisite variety 175
research agenda for developmental work
    198–9

resilience engineering approach
    characteristics of 21–3
    contribution to resilience by humans
        76–8
    increasing resilience of normal
        practices 44–5
    in real situations 23–5
risk control, contribution to by humans
    76–8
risk management 131
risk(s) 130
    external and internal 131–3, 135–7
    historical context of in aviation
        industry 174
    vulnerability to of high-3 work
        173–4
Roberts, P. 83
Robinson, M. 101
Rochlin, G.I. 187
Rogalski, J. 134, 185, 204
Rørøs line accident. *see* Åsta accident on
    Rørøs line
Rosness, R. 14
Rouse, W.B. 140

S

safety instrument system, design process
    for 162–7, *164*
safety management 134, 140–1, 142–5
Salo, I. 21
SAMU. *see* emergency medical centre
Schön, D. 157
Scribner, S. 155
sequencing in air traffic control (ATC)
    191–2
sequential models 20–1
Service d'Aide Médicalisée d'Urgence
    (SAMU). *see* emergency medical
    centre
shared mental models 140
simultaneity in air traffic control (ATC)
    191–2
situated action 100, 154
Skjerve, A.B. 77
Sleipner A incident 81
Snook, S.A. 14, 55
Snorre A incident 82, 95–6
sociotechnical systems, drift in 54–5

standardization of healthcare 109n2
stress in systems. *see* systemic stress
structural change and problem-solving 71
Suchman, L.A. 24, 100, 154, 184
sufficiency 133
Super Puma helicopter incident 81
Svedung, I. 20, 69
Svensson, O. 21
systematic working as essential in investigations 19
systemic accident analysis method 26–7
systemic models 20
systemic stress 15
    as theme for developmental work 200–1
    Finnish piloting 42
systems, vulnerabilities in 82

T
task multiplicity 187–8
technologically mediated navigation 37–8, **39, 40**
tempo of work in ATC 180
temporal dimension of air traffic control (ATC) 177, **178,** 178–83, *179*
temporality in negotiating action 102
Tenerife airport jumbo jet collision 55
tension in systems. *see* systemic stress
thinking, higher-order 183
tight coupling 54, 55
traditional navigation 37–8, **39, 40**
trajectories in negotiating action 102
Tretton accident 62n1
Turner, Barry 70, 71, 92–3

U
Upton, D. 83

V
values and incremental decision-making 72
Vaughan, D. 73
Vicente, K.J. 7, 153, 154
Von Cranach, M. 134
vulnerability 92
    definition 82
    emergence of 14
    to risk of high-3 work 173–4

W
Wackers, G. 199
war stories in air traffic control (ATC) 187
watchdogs, need for with incremental decision-making 72–3
Weick, Karl 22, 55, 175, 192
Woods, D. 22
work
    analysis of 161–2
    new generation of models for 153
work organization
    refocusing on 174
    situated action 100
    tension in 100
work processes, ordering and coordinating 92–5
Wreathall, J. 23

Y
Yergin, D. 83